FLORA OF TROPICAL EAST AFRICA

URTICACEAE

I. FRIIS

University of Copenhagen, Denmark

Herbs, shrubs, lianas or small trees, monoecious, dioecious or rarely polygamous, some genera with stinging hairs, but also frequently with stiff non-stinging hairs, which may be curled or sharply hooked; usually with punctiform, elongated or linear cystoliths in epidermal cells. Leaves alternate or opposite, sometimes anisophyllous, petiolate or sessile; stipules usually present, lateral or often intrapetiolar, often fused; lamina simple or 3–5(–7)-lobed, margin entire or variously dentate, usually with 3 subequal nerves from the base, the basal pair of nerves reaching towards the leaf-apex, the other lateral nerves usually weaker and shorter, cystoliths usually clearly visible in the epidermis of the upper surface. Inflorescences very varied, mostly pedunculate, lax or condensed racemes, often with flowers in small cymose glomerules, or sessile and condensed cymes in leaf-axils, partial inflorescences often subtended by involucral bracts, the inflorescence-axis sometimes flattened into a disc-shaped, ± fleshy receptacle. Flowers minute, unisexual or rarely bisexual, actinomorphic or (especially in ♀ flowers) zygomorphic, with single whorl of tepals or rarely the ♀ flowers naked, pedicel often articulated below perianth. Male flowers with (1–)2–5(–6) tepals, free or connate in lower half; stamens equalling tepals in number or solitary, always inflexed in bud; rudimentary ovary often present. Female flowers with 3–5 tepals, free or ± completely united, often very unequal, often accrescent after pollination, rarely rudimentary or completely absent; staminodes, if present, rudimentary or scale-like and important for the dispersal of the fruit (ejecting the achene when reflexing); ovary superior, usually laterally compressed, often oblique or asymmetrical, unilocular, unicarpellate, placentation basal, with 1 erect ovule; stigma capitate, penicillate or filiform, linear. Fruit an achene, sometimes enclosed by persistent accrescent perianth which may become fleshy.

About 50 genera and approximately 1000 species; almost cosmopolitan but most numerous in the tropics. Most genera and species occur in humid habitats, e.g. on the forest floor or in most places at forest margins, often along streams, but some genera are adapted to arid areas, e.g. *Obetia* and *Forsskaolea*.

The genera may be referred to in the following conventional tribes, which are useful as far as they class together genera with various sets of common features, though it is not certain that the rank and delimitation of the tribes reflect the phylogeny of the family.

Tribe **Urticeae** (*Urereae* (Gaudich.) Wedd.). Herbs, shrubs or trees usually armed with stinging hairs; stipules present. Male flowers 4–5-merous, usually with rudimentary ovary. Female flowers usually with 4 tepals; staminodes absent. Genera 1–5.

Tribe **Procrideae** *Wedd.* (*Elatostemeae* Gaudich., sine stat. def.). Usually juicy, moisture-loving herbs without stinging hairs; stipules present. Male flowers (2–)4–5-merous, without or rarely with rudimentary ovary. Female flowers mostly with 3 ± fused tepals (perianth-lobes) and with 3 scale-like inflexed staminodes which forcefully eject the achene at maturity; stigma penicillate. Genera 6–8.

Tribe **Boehmerieae** (*Gaudich.*) *Wedd.* Herbs or more often shrubs or small trees; stipules present. Male flowers 4–5-merous, with or without rudimentary ovary. Female flowers mostly with tubular perianth ± free from the ovary (in genera outside the Flora area also sometimes without perianth); stigma almost always filiform. Genera 9,10.

Tribe **Parietarieae** (*Gaudich.*) *Wedd.* Annual herbs, outside Flora area also perennial herbs and shrubs; stipules mostly absent. Male flowers 3-, 4- or 5-merous, with or without rudimentary ovary. Female flowers with tubular or 4-lobed or partite perianth; staminodes absent; stigma linear or penicillate. Bisexual flowers sometimes present. Genus 11.

Tribe **Forsskaoleae** (*Gaudich.*) *Wedd.* Annual or perennial herbs; stipules present. Inflorescences sometimes reduced to a single flower, often enclosed in an involucre. Male flowers ± boat-shaped, with 1 tepal and 1 stamen. Female flowers with tubular perianth (the nature of the perianth can be disputed, it may in some cases be considered a reduced involucre) or naked; stigma filiform or penicillate. Genera 12–15.

NOTE.   The genus *Lecanthus* Wedd. has been indicated as occurring in E. Zaire, Uganda, Kenya, Tanzania, and Malawi by Letouzey in Fl. Cameroun 8: 183 (1968). While this is not unlikely, as the species *Lecanthus peduncularis* (Royle) Wedd. grows in Ethiopia and Cameroon, it has not been possible to find any material to confirm Letouzey's statement. The plant is a small annual occurring in moist places, usually associated with rocks, and characterized by having opposite leaves as in *Pilea* and disc-shaped inflorescences like those in *Elatostema*. It should be looked for in East Africa.

1. Plants with stinging hairs, at least on the inflorescences
    and petioles 2
    Plants completely without stinging hairs 6
2. Leaves opposite . . . . **1. Urtica**
    Leaves alternate 3
3. Stipules free, lateral; small tree with terminal rosettes
    of leaves . . . . **3. Obetia**
    Stipules fused, intrapetiolar; woody climbers or erect,
    short-lived herbs, leaves ± scattered along the stems 4
4. Woody climbers with adventitious roots from the climbing
    stems, stinging hairs mostly restricted to petioles
    and/or inflorescences; ♀ perianth becoming fleshy in
    fruit . . . . **2. Urera**
    Erect perennial herbs or small shrubs, not climbing and
    without adventitious or axillary roots, at least from
    erect stems; stinging hairs usually common on most
    aerial parts; ♀ perianth always membranous in fruit 5
5. Female perianths with 3 almost completely fused tepals,
    forming a 1-sided cover on the ovary or achene;
    longest stinging hairs almost always more than 5 mm.
    long . . . . **4. Girardinia**
    Female perianths with 4 free, very unequal tepals (one
    large and one small pair), ± covering the sides of the
    ovary or achene; stinging hairs shorter than 5 mm.   **5. Laportea**
6. Leaves always opposite 7
    At least some of the leaves alternate (sometimes also some
    leaves whorled in 10. *Pouzolzia*) 11
7. Stipules fused, intrapetiolar; the 2 leaves of a pair often
    very different in size and shape 8
    Stipules free, lateral; the 2 leaves of a pair always equal or
    subequal in size and shape 9
8. Female flowers with 3-lobed perianth (lobes different or
    similar in size and shape) and 3 large scale-like
    staminodes, acting as ejectors of the achene . . .   **6. Pilea**
    Female flowers with 4–5 very short free, subequal tepals;
    staminodes reduced or absent; achene not ejected,
    dispersed with perianth . . . . **8. Procris**
9. Male flowers with 4 stamens; inflorescences pendent
    interrupted spikes . . . . **9. Boehmeria**
    Male flowers with only 1 stamen; inflorescences axillary
    clusters or pedunculate few-flowered heads 10
10. Bisexual or ♂ inflorescences enclosed by a common
    involucre; ♀ inflorescence with clusters of small, 1(–2)-
    flowered involucres . . . . **13. Droguetia**
    Inflorescences without involucre . . . . **14. Australina**
11. Some leaves whorled . . . . **10. Pouzolzia**
    Leaves never whorled 12
12. Stipules absent . . . . **11. Párietaria**
    Stipules present at least at the youngest leaves 13

13. Leaves distichous, sessile or subsessile, markedly asymmetrical relative to the midnerve . . . . . **7. Elatostema**
Leaves mostly spirally arranged, petiolate, symmetrical (or almost so) relative to the midnerve 14
14. Flowers enclosed in a campanulate involucre ± filled with a lanate indumentum; fruit enclosed in and dispersed with the involucre 15
Flowers and fruits not enclosed in an involucre 16
15. Involucral bracts almost completely free . . . . . . **12. Forsskaolea**
Involucral bracts fused almost to the apex . . . . . **13. Droguetia**
16. Inflorescence a lax panicle or an interrupted spike 17
Inflorescence a dense axillary cluster 18
17. Woody climber; achene at maturity enclosed in the fleshy persistent perianth . . . . **2. Urera**
Erect herbaceous or slightly woody plant; ♀ perianth with free tepals, membranous in fruit . . . . **5. Laportea**
18. Male flowers with 4–5 stamens; achene at least partly free from the persistent perianth . . . . **10. Pouzolzia***
Male flowers with 1 stamen; achene without or completely fused with the perianth . . . . **15. Didymodoxa**

1. URTICA

L., Sp. Pl.: 983 (1753) & Gen. Pl., ed. 5: 423 (1754); G.P. 3: 381 (1880); Engl. in E. & P. Pf. 3(1): 104 (1888); G.F.P. 2: 181 (1967)

Annual or perennial herbs, monoecious or dioecious. Stems with stinging and sometimes stiff hairs. Leaves opposite, petiolate, simple; stipules lateral, free, or interpetiolar, fused; cystoliths punctiform. Inflorescences axillary, mostly shortly pedunculate, often paired in each leaf-axil, bisexual or unisexual lax cymose panicles. Flowers unisexual, 4-merous, usually clustered in small cymose glomerules. Male flowers: tepals free, in 2 subequal pairs; rudimentary ovary present. Female flowers: tepals free, in 2 ± unequal pairs; staminodes absent; ovary ovoid, laterally compressed, symmetrical; stigma sessile, penicillate. Achene enclosed in or released from the persistent perianth, lenticular with a ± raised central area on each face.

About 50 species, mostly in the northern temperate regions, with a few in the tropics and southern temperate regions. Five species are found in Africa, in the mountains of Ethiopia and East Africa, and in South Africa; three are indigenous and the other two are introduced weeds. In East Africa one indigenous and one introduced species.

Stout perennial . . . . **1. U. massaica**
Slender annual . . . . **2. U. urens**

1. U. massaica *Mildbr.* in N.B.G.B. 8: 275 (1923); F.D.O.-A. 2: 116 & Appendix: 10, t. 13/A–E (1932); Hauman in F.C.B. 1: 180 (1948); A.V.P: 72 (1957); U.K.W.F.: 321 (1974); Troupin, Fl. Rwanda 1: 153, fig. 30/3A–C (1978). Type: Tanzania, Kilimanjaro, Olmolog, *Endlich* 170 (B, holo.†)

Erect perennial dioecious herb up to 2 m. tall, with little-branched ± quadrangular stems, forming loose clumps from a creeping rhizome; all parts of plant with fiercely stinging hairs 1.5–2 mm. long, raised on protuberances ± 1 mm. high, otherwise glabrous to puberulous. Leaves: stipules fused, interpetiolar, (0.8–)1–2 cm. long, 0.4–1 cm. wide, with rounded apex and broadly cuneate to cordate base, chartaceous, brown; petioles 1.5–4.5 cm. long; lamina ovate, 7–13 cm. long, 6–10.5 cm. wide, base cordate, margin serrate or usually double serrate, with numerous teeth on each side, apex acute to shortly acuminate; lateral nerves 6–9 pairs, lower 2–3 pairs running into subcordate leaf-base, all pairs extending to leaf-margin and linked by a fine net of anastomosing nerves; upper surface with stinging hairs and sometimes also a fine pubescence, cystoliths very

---

* *Boehmeria nivea*, reportedly occasionally cultivated in East Africa, would key out here.

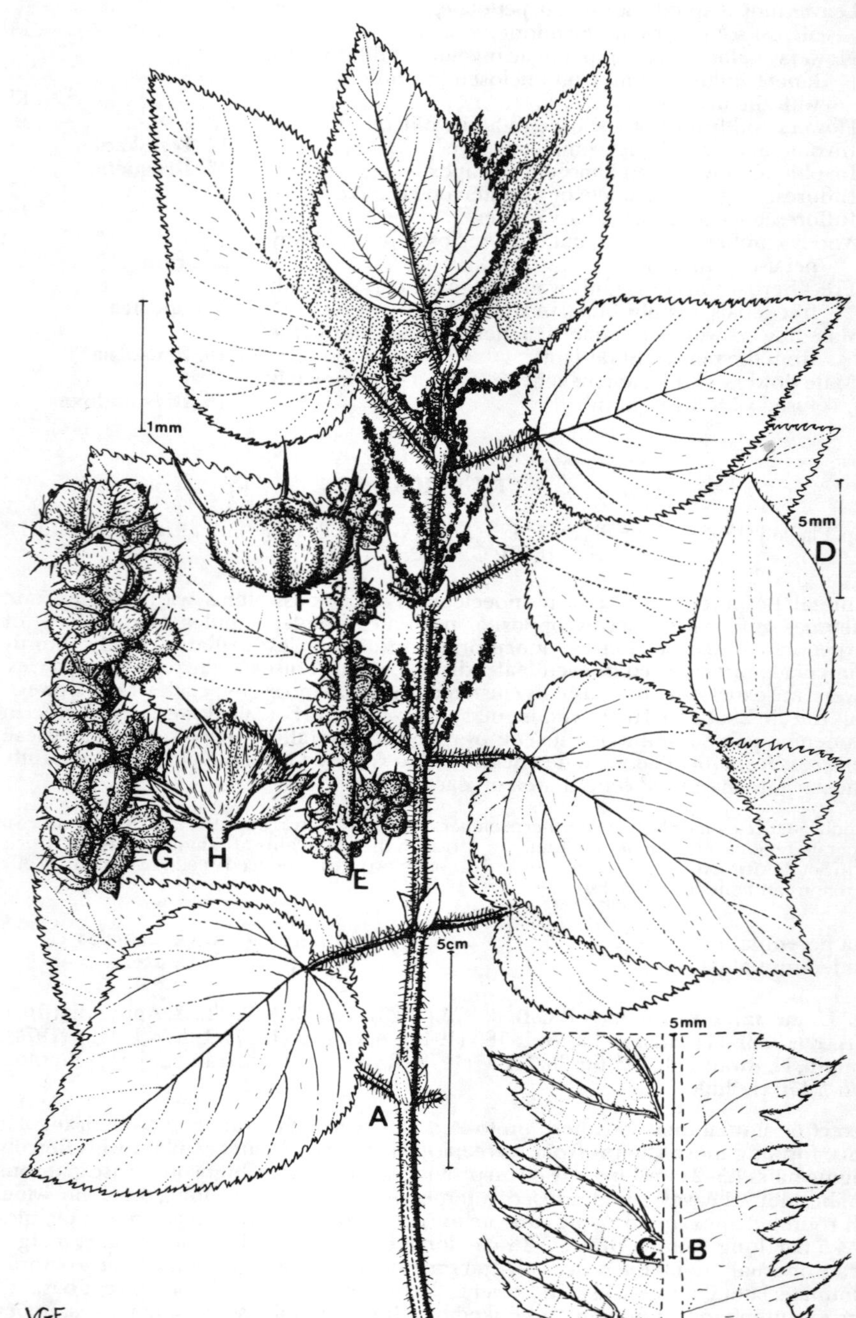

FIG. 1. *URTICA MASSAICA* — **A**, flowering stem; **B**, detail of leaf from above; **C**, detail of leaf from below; **D**, stipule; **E**, detail of male inflorescence; **F**, male flower in bud; **G**, detail of female inflorescence; **H**, female flower. A–D, from *Kibui* 52; E, F, from *Kanuri* 13634; G, H, from *Bally* 2346. Drawn by Victoria C. Friis.

numerous, punctiform, lower surface with stinging hairs along the nerves. Inflorescences paired in the axils of the upper leaves, spike-like, 2.5–4.5 cm. long, with the flowers in small, densely cymose, bracted glomerules, sometimes the inflorescence with a few side-branches near the base. Male flowers on pedicels 0.8–1.5 mm. long; perianth ± 1 mm. in diameter, puberulous to pubescent, occasionally with a few stinging hairs, depressed globose in bud. Female flowers sessile; perianth ± 1 mm. long, pubescent, with numerous stinging hairs on the large pair of tepals. Achene flattened, ovoid, ± 1 mm. long, 0.8 mm. wide, white, long enclosed in the persistent perianth but dispersed without it. Fig. 1.

UGANDA. Kigezi District: Kachwekano Farm, July 1949, *Purseglove* 3031! & Kabale, July 1945, *A. S. Thomas* 4306!
KENYA. Naivasha District: Aberdare Mts., Chebuswa [Eland Hill], Oct. 1970, *Mabberley* 393!; S. Nyeri District: Kahuroini, Githi, Oct. 1964, *Kibui* 52!; Masai District: Nasampolai [Enesambulai] valley, July 1969, *Greenway & Kanuri* 13634!
TANZANIA. Mbulu District: Mt. Hanang, May 1962, *Polhill & Paulo* 2301!; Arusha District: Ngurdoto Crater, Oct. 1965, *Greenway & Kanuri* 11983!; Lushoto District: W. Usambara Mts., Makuyuni, June 1935, *Koritschoner* 883!
DISTR. U 2,4; K 1, 3, 4, 6; T 2, 3; E. Zaire, Rwanda, Burundi
HAB. Clearings and natural open glades in upland rain-forest and moist evergreen bushland, often near human dwellings, especially where cattle are kept; (400–)1500–3250 m.

SYN. [*U. simensis* sensu Rendle in F.T.A. 6(2): 243 (1917), *non* Steud. (1850)]

NOTE. Closely related to the Ethiopian *U. simensis* Steud. and suggested to be conspecific by Rendle, but it seems perfectly feasible to distinguish the two taxa on several characters. In *U. simensis* the stipules are 0.6–1(–1.3) cm. long, the leaves usually narrower and with simple serration, the ♂ flowers ± 2 mm. in diameter, and the achenes ± 2 mm. long.
    The sting of *U. massaica* is recorded as painful but wears off after a few minutes. The leaves may be boiled and eaten, or used young and uncooked as a spinach, as is the case with many species of *Urtica* throughout the world. Reports that the roots are toxic (*Koritschoner* 883), or that the plant is useful to keep the lions out (*B. D. Burtt* 4286), are more fanciful.
    Peter, F.D.O.-A. 2 (1932), provided a new validating description and new types, presumably overlooking the earlier publication by Mildbraed.

2. U. urens *L.*, Sp. Pl.: 984 (1753); Rendle in F.T.A. 6(2): 242 (1917). Type: Sweden, collector unknown, *Linnaean Herbarium* 1111.5 (LINN, lecto.!)

Erect, ascending or procumbent annual herb up to 0.5 m. high, sometimes profusely branched from base, monoecious. Stems with stiff hairs when young, later glabrescent; stinging hairs dense to sparse. Leaves: stipules narrowly lanceolate, ± 1.5 mm. long, 0.5 mm. wide, acute, pubescent; petiole (0.7–)3–5 cm. long, with stinging and stiff hairs; lamina elliptic, 2.5–4.5 cm. long, 1.5–2.8 cm. wide, base cuneate, margin with 9–13 long narrow teeth on each side, the teeth often 3-lobed, especially in the lower half; basal pair of lateral nerves reaching the upper $\frac{1}{3}$ of lamina; upper surface with scattered stinging hairs, lower surface with stinging hairs on the nerves, stiff hairs absent or sometimes a few present on young leaves. Inflorescences dense at first, becoming lax, cylindrical, spike-like, 0.5–2.5 cm. long, 0.3–0.75 cm. wide, bisexual, with ♂ and ♀ flowers mixed, frequently with both stinging and stiff hairs; peduncle short. Male flowers on pedicels ± 0.5 mm. long; perianth ± 1 mm. in diameter. Female flowers on pedicels 0.5–1 mm. long; perianth 1.5–2 mm. long, the 2 large tepals each with a stinging hair; ovary enclosed in the perianth, with only the penicillate stigma protruding. Achene laterally compressed, smooth or minutely punctate, pale ochre.

KENYA. Naivasha District: S. Kinangop, July 1965, *Gillett* 16765!
TANZANIA. W. Usambara Mts., Gare, Sept. 1945, *Greenway* 7536!
DISTR. K 3; T 3; widespread through all temperate parts of the world and in the highlands of the tropics, presumably originating in the temperate region of the Old World
HAB. In cultivations and on old hut sites; 1800–2500 m.

2. URERA

Gaudich. in Freyc., Voy. Monde Bot.: 496 (1830); G.P. 3: 383 (1880); Engl. in E. & P. Pf. 3(1): 105 (1888); G.F.P. 2: 181 (1967)

Woody climbers, often fixed to the substrate by adventitious axillary roots from the stems, dioecious, with stinging hairs on petioles and inflorescences, stiff hairs absent;

young stems and petioles often with protuberances bearing stinging hairs. Leaves alternate, petiolate, simple; stipules intrapetiolar, fused, with free tips; cystoliths punctiform or elongated. Inflorescences of lax axillary racemes with the flowers in small cymose glomerules. Male flowers 4–5-merous, with rudimentary ovary. Female flowers with 4 somewhat unequal tepals fused in the lower part or with a ± completely tubular perianth bearing ± 4 indistinct teeth at the opening; staminodes absent; ovary ovoid, ± completely enclosed in the perianth; stigma almost sessile, penicillate. Fruit an achene, enclosed in the persistent perianth which enlarges and becomes fleshy and red in fruit.

About 40 species, widespread in tropical Africa, Madagascar, tropical America and Pacific Is. The genus is rather heterogeneous in perianth morphology.

1. Leaf-margin entire . . . . *1. U. trinervis*
   Leaf-margin serrate or crenate 2
2. Leaves with serrate margin, always densely pubescent to white-felted beneath; stinging hairs on the young stems and petioles, raised on protuberances (1–)2–3 mm. high; ♀ flowers with tubular perianth . . . *2. U. sansibarica*
   Leaves with crenate margin, puberulous, rarely pubescent beneath; stinging hairs on young stems and petioles never raised on protuberances more than 0.5 mm. high; ♀ flowers with 2 large and 2 small subcircular lobes, free almost to base . . . . *3. U. hypselodendron*

1. U. trinervis (*Hochst.*) *Friis & Immelman* in Nordic Journ. Bot. 7: 126 (1987). Type: South Africa, Natal, Umlaas R., *Krauss* 1267 (TUB, holo.!, C, K, PRE, photo.)

Robust climber to 10 m. or more. Stems softly woody, up to 10 cm. in diameter at base; bark grey to brownish black, longitudinally striate, glabrous, rarely pubescent, with large leaf scars on young branches but very rarely with protuberances; sap copious, clear; pith wide, spongy, or stems hollow in centre. Leaves: stipules brown, fused almost to the apex, 6–10 mm. long, pubescent on the midnerves to subsericeous outside, glabrous inside; petioles 2.5–6 cm. long, glabrous to pubescent, rarely with a few stinging hairs, which are sessile or raised on protuberances up to 0.5 mm. high; lamina coriaceous, elliptic to ovate, rarely obovate, (4–)6–12 cm. long, 3.5–8 cm. wide, base cuneate, truncate or rounded, rarely subcordate, margin entire, apex acuminate to caudate; lateral nerves 2–3(–4) pairs, the basal pair extending into upper ⅓ of lamina, all interconnected by scalariform tertiary nerves; upper surface glabrous, rarely with a few scattered stiff hairs, numerous elongated cystoliths present, lower surface glabrous, occasionally pubescent or with scattered, stiff and perhaps stinging hairs, especially on the lateral nerves, rarely with stinging hairs raised on protuberances up to 0.8 mm. high. Male inflorescence lax, paniculate, ± 6.5 cm. long, with flowers densely clustered at intervals along the axes; peduncle mostly with a few stinging hairs; ♀ inflorescence lax, paniculate, ± 2 cm. long, with small cymose clusters of flowers, peduncle with stinging hairs, especially around the clusters of flowers. Male flowers on pedicels up to 1 mm. long, 4-merous, glabrous, perianth 1–1.75 mm. in diameter. Female flowers sessile; perianth cylindrical, constricted at apex, with 4 blunt but clearly marked teeth, glabrous, 1–1.5 mm. long. Achene glabrous, 1.5–2 mm. long, without markings, enclosed in the persistent, accrescent, fleshy, orange perianth. Fig. 2/H–K, p. 8.

UGANDA. Ankole District: Lutoto, July 1939, *Purseglove* 833!; Masaka District: Jubiya Forest near Luunga, May 1969, *Lye* 2969!; Mengo District: Kasa Forest, Dec. 1949, *Dawkins* 477!
KENYA. Nandi District: Kaimosi, SW. of Yala Bridge, Mar. 1965, *Gillett* 16705!; Masai District: Mzima Springs, June 1953, *Bally* 8977!; Teita District: Mt. Kasigau, Feb. 1971, *Faden et al.* 71/116!
TANZANIA. Bukoba District: Rubare Forest Reserve, Feb. 1958, *Procter* 808!; E. Usambara Mts., Amani, Jan. 1967, *Harris* 611!; Pangani District: S. bank of Pangani R., Hale–Makinyumbe, July 1953, *Drummond & Hemsley* 3124!
DISTR. U 2,4; K 3, ?4, 5–7; T 1–4, 6; widely distributed in tropical Africa from Ghana through Central Africa to SW. Ethiopia, through East Africa to Angola and South Africa (Natal, where restricted to near coast) as far south as Port St. Johns; apparently absent from Malawi and Mozambique; also on Madagascar
HAB. Lowland rain-forest, riverine forest or swamp forest, especially along margins and in clearings, sometimes sprawling over rocks or among boulders; 200–1600 m.

SYN. *Elatostema trinerve* Hochst. in Flora 28: 88 (1845)

Urera cameroonensis Wedd. in DC., Prodr. 16(1): 97 (1869); Rendle in F.T.A. 6(2): 261 (1917);
   Hauman in F.C.B. 1: 185 (1948); F.W.T.A., ed. 2, 1: 618 (1958); Letouzey in Fl. Cameroun 8: 76
   (1968); U.K.W.F.: 321 (1974); Troupin, Fl. Rwanda 1: 153, fig. 30/1A,B (1978); Friis in Nordic
   Journ. Bot. 5: 549 (1986). Type: Cameroon, Mt. Cameroon, *Mann* 2173 (K, holo.!)
U. arborea De Wild. & Th. Dur. in B.S.B.B. 38: 52 (1900). Type: Zaire, Mbandaka [Coquilhatville],
   Dewèvre 618 (BR, holo.!)
U. laurentii De Wild., Miss. Laurent 1: 72, t. 20 (1905). Type: Zaire, Ubangui, Imese, *E. & M.
   Laurent* (BR, holo.!)
U. gilletii De Wild. in Ann. Mus. Congo, Bot., sér. 5,1: 240 (1906). Type: Zaire, Bas Congo,
   Kisantu, *Gillet* 2312 (BR, holo.!)
U. woodii N.E. Br. in K.B. 1911: 96 (1911) & in Fl. Cap. 5(2): 549 (1925). Types: South Africa, Natal,
   Port St. Johns, *Pegler* 1533 (K, syn.!, BM, BOL, isosyn.!) & near Umzinyati Falls, *J. M. Wood* 1803
   (K, syn.!) & without precise locality, *Sanderson* 594 (K, syn.!)
U. usambarensis Rendle in J.B. 54: 370 (1916) & in F.T.A. 6(2): 263 (1917). Type: Tanzania, E.
   Usambara Mts., Derema, *Scheffler* 196 (K, syn.!) & *Volkens* 122 (BM, syn.!, K, isosyn.!)
U. cameroonensis Wedd. var. *laurentii* (De Wild.) Rendle in F.T.A. 6(2): 262 (1917)
U. acuminata (Poir.) Decne. var. *cameroonensis* (Wedd.) Leandri in Fl. Madag. 56: 24 (1965), excl.
   spec. cit.

NOTE.   A rather distinct group of specimens has been collected from the W. and S. banks of Lake
   Victoria (**U** 4: Namanve, *Eggeling* 778; **T**1: Kibura, *Tanner* 574; Shinyanga, *Koritschoner* 3001; Speke
   Gulf, *B. D. Burtt* 2487; Kigongo, *Tanner* 1178; Bukoba, *Procter* 703). These specimens are generally
   more pubescent than normal, and have stiff, stinging hairs on the leaves above and on the nerves
   beneath, sometimes even on the stems, and usually an obovate lamina, frequently with more than
   two pairs of lateral nerves above the prominent basal pair. The deviating form occurs
   sympatrically with the normal form, and could therefore be considered a variety of local
   distribution. However, in view of the total variation of this very widespread and variable species, it
   has been decided not to accord any formal name to the form from the Lake Victoria area.
      The species approaches the closely related *U. thonneri* De Wild. & Th. Dur. from Cameroon,
   Zaire and Angola, from which it differs in details of leaf morphology, and *U. acuminata* (Poir.)
   Decne. from the Mascarenes, from which it differs by its constantly 4-merous ♂ flowers.
      Used as string or rope (*Akeroyd & Mayuga* 60); the species is frequently used in Zaire as a
   vegetable and for its fibre, for which reason it is sometimes planted.

2. U. sansibarica *Engl.*, P.O.A. C: 162 (1895); Rendle in F.T.A. 6(2): 263 (1917); Friis in
Nordic Journ. Bot. 5: 549 (1986). Type: Tanzania, Zanzibar I., Kidoti, *Hildebrandt* 1039 (B,
holo.†, BM, K, iso.!)

Liana climbing to 3 m. or perhaps more. Stems with reddish brown, often peeling bark,
usually with numerous protuberances up to 5 mm. high, crowned by stinging hairs;
younger stems densely pubescent to hirtellous, older stems glabrescent. Leaves: stipules
fused for at least ⅔ of the length, 0.5–1 cm. long, densely pubescent outside, glabrous
inside; petiole 2.5–11.5 cm. long, usually pubescent and densely covered by
protuberances with stinging hairs; lamina herbaceous, dark green, often drying almost
black, with paler underside because of dense indumentum, ovate to elliptic, 7.5–15 cm.
long, 5–10.5 cm. wide, base cordate to subcordate or truncate, margin serrate with 25–35
teeth on each side, apex bluntly acute to acuminate; lateral nerves 4–5 pairs, basal pair
extending into upper ½–⅓; upper surface subscabrous to smooth, glabrous or with a few
short, stiff hairs, cystoliths punctiform, rarely somewhat elongated, lower surface with
stinging hairs on the nerves and usually densely white-felted between the nerves or on
the fine reticulation of the ultimate nerves. Male inflorescences on 1–2 cm. long
peduncles, up to 10 cm. in diameter, paniculate, with glabrescent to pubescent axes;
flowers in small clusters at the branching-points and along the axes; ♀ inflorescence
sessile or on peduncles up to 2 cm. long, pubescent, up to 5 cm. in diameter, with flowers
in clusters surrounded by groups of stinging hairs. Male flowers on pedicels 1–1.5 mm.
long, articulated below perianth, 5-merous, perianth ± 1 mm. in diameter. Female flowers
sessile; perianth tubular, with 4 teeth at apex; ovary enclosed except for penicillate stigma.
Achene compressed, ± 1.5 mm. long, enclosed in the persistent, red, fleshy, accrescent
perianth. Fig. 2/L–N, p. 8.

KENYA.   Kwale District: Shimba Hills, Sheldrick Falls, Apr. 1968, *Magogo & Glover* 613! & Mrima Hill,
   Jan. 1964, *Verdcourt* 3939C!; Kilifi District: Chasimba, June 1973, *Musyoki & Hansen* 954!
TANZANIA.   Lushoto District: Kisiwani Forest, Sept. 1961, *Ali Omare in Richards* 15329!; Handeni
   District: Kwamkono [Kwa Mkono], Feb. 1980, *Archbold* 2706!; Tanga District: Kange, June 1958,
   Faulkner 2154!
DISTR.   **K** 7; **T** 3, 6; **Z**; NE. Mozambique
HAB.   Lowland rain-forest, often on limestone, frequently in glades and climbing over rocks;
   5–900 m.

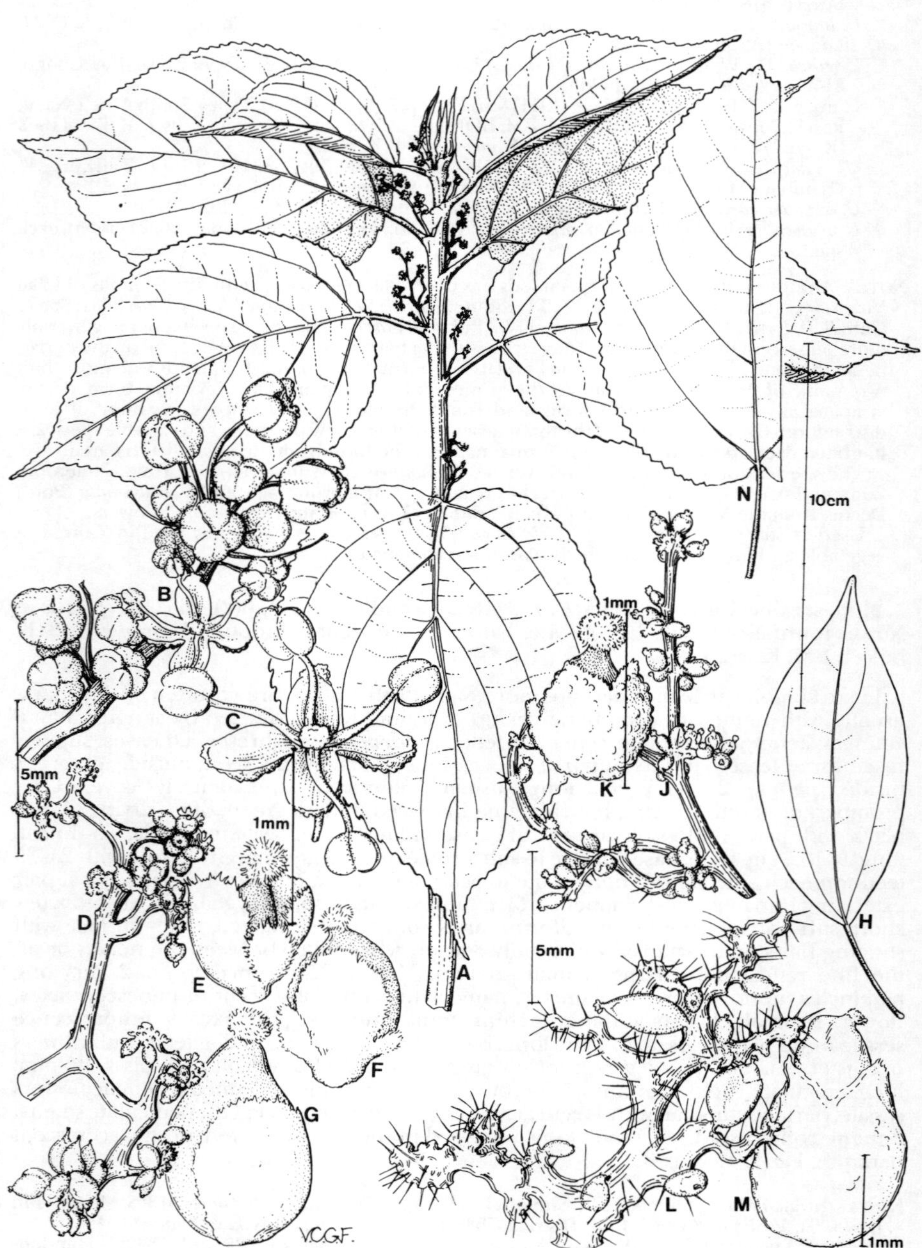

FIG. 2. *URERA HYPSELODENDRON* — **A**, flowering stem; **B**, detail of male inflorescence; **C**, male flower; **D**, detail of female inflorescence; **E**, young female flower; **F**, old female flower; **G**, fruit with persisting fleshy perianth. *U. TRINERVIS* — **H**, leaf; **J**, detail of female inflorescence; **K**, old female flower with persisting, fleshy perianth. *U. SANSIBARICA* — **L**, part of female inflorescence; **M**, old female flower with persisting, fleshy perianth; **N**, leaf. Drawn by Victoria C. Friis.

SYN. *U. fischeri* Engl., P.O.A. C: 162 (1895); Rendle in F.T.A. 6(2): 264 (1917). Type: ? Tanzania
　　[Ostafrika], without precise locality, *Fischer* 117 (B, holo.†)
　　U. braunii Engl. in V.E. 3(1): 53 (1915). Type from Tanzania, Usambara and Pare Mts., not
　　precisely indicated; according to Peter in F.D.O.-A.2: 117 (1932), a specimen from the
　　Useguha area, collected at Hale on the Pangani R. by *Braun* in *Herb. Amani* 1520 (EA, 3 sheets)
　　may represent original material

NOTE. Although no type material has been specified in the protologues of *U. braunii* or *U. fischeri*,
　　the descriptions of the two taxa point to synonymy with *U. sansibarensis*. However, Peter, in
　　F.D.O.-A. 2: 117 & 119 (1932), cited specimens of *U. braunii* from W. Usambara Mts., Mt. Meru, and
　　Uluguru Mts. and specimens of *U. fischeri* from E. Usambara Mts. and Tanga District, Udigo; the
　　specimens from W. Usambaras and Mt. Meru could be taken as an indication that *U. braunii* is
　　conspecific with *U. trinervis*, while the *Braun* belongs here.
　　　　On a specimen from the Uluguru Mts. (*E. M. Bruce* 1022) it is recorded that the fibres of this
　　species are used for tying.

3. U. hypselodendron (*A. Rich.*) *Wedd.* in Ann. Sci. Nat., sér. 3, 18: 203 (1852); Rendle in
F.T.A. 6(2): 255 (1917); Hauman in F.C.B. 1: 181 (1948), excl. tab.; U.K.W.F.: 321 (1974);
Troupin, Fl. Rwanda 1: 153 (1978), excl. fig. 30/2A–C; Friis in Nordic Journ. Bot. 5: 551
(1985). Types: Ethiopia, Semien, Endchedkap to Shouata [Choada], *Schimper* 1136 (P,
syn.!, BM, K, isosyn.!) & Shire [Chire], *Quartin-Dillon* (P, syn.)

Liana climbing to a height of 25 m. or more. Stems with numerous adventitious roots;
bark of young stems reddish-brown, pubescent to glabrescent, without protuberances
and stinging hairs but with numerous large leaf-scars; older stems with glabrous, smooth
or striate, reddish-brown bark, usually not peeling; sap abundant, clear. Leaves: stipules
lanceolate, fused for at least ⅔ of their length, 0.7–1.4 cm. long, 0.3–0.5 cm. wide,
caducous, dark brown, puberulous outside, especially on the nerves, often ciliate; petiole
1.5–6.5 cm. long, glabrous to pubescent, very rarely with short protuberances and stinging
hairs; lamina dark green, drying brownish black, ovate, 7–16.5 cm. long, 4–11.5(–15) cm.
wide, base subcordate to cordate, margin crenate-dentate, with more than 35 crenations
on each side, apex acuminate, rarely acute; lateral nerves 5–7 pairs, basal pair extending
to upper ⅓ of lamina, tertiary nerves and nerves of higher order clearly marked, parallel,
forming a scalariform pattern; upper surface glabrous, occasionally with a few stiff hairs,
cystoliths punctiform to slightly elongated, lower surface glabrescent to puberulous,
especially on the nerves, rarely with the whole surface pubescent, or with a few stinging
hairs on the nerves. Male inflorescences sessile or on peduncles up to ± 1 cm. long, lax,
irregularly paniculate, up to ± 15 cm. in diameter, axes glabrescent to pubescent,
sometimes with a few stinging hairs; ♀ inflorescences sessile or on peduncles up to 2 cm.
long, up to ± 8 cm. in diameter, glabrescent to pubescent, occasionally with protuberances
1–3 mm. high, bearing stinging hairs. Male flowers in clusters up to ± 0.5 cm. in diameter,
on minutely pubescent pedicels 1–2.5 mm. long, articulated below the perianth, 4-
merous, perianth ± 2 mm. in diameter. Female flowers sessile to subsessile, densely
clustered, on pedicels ± 1 mm. long; perianth-segments 4, basally fused, very unequal, 2
outer short, 2 inner subcircular, almost as long as the ovary; stigma penicillate,
protruding. Achene 1–1.5 mm. long, compressed, slightly oblique, brown, minutely
granulate, enclosed by the 2 orange-red, fleshy, accrescent perianth-lobes. Fig. 2/A–G.

UGANDA. Kigezi District: Rutenga, Apr. 1948, *Purseglove* 2630! & Impenetrable Forest, June 1938,
　　Eggeling 3687!; Mbale District: Mt. Nkokonjeru, Mar. 1927, *Snowden* 1042!
KENYA. NE. Elgon, May 1958, *Tweedie* 1557!; Nakuru District: along Kericho road, SW. of Londiani,
　　Nov. 1962, *Perdue & Kibuwa* 9128!; Machakos District: Donyo Sabuk [Sapuk], Jan. 1964, *Napper et al.*
　　1717!
TANZANIA. Mbulu District: Hanang Forest Reserve, Oct. 1968, *Carmichael* 1567!; Mbeya District:
　　Poroto Mts., Livingstone Forest Reserve, Sept. 1970, *Thulin & Mhoro* 1230!; Iringa District:
　　Mufindi, Kilima, Mar. 1962, *Polhill & Paulo* 1805!
DISTR. **U** 1–3; **K** 3–6; **T** 2, 3, 5–8; Sudan (Imatong Mts.), Ethiopia, E. Zaire, Rwanda, Burundi, Malawi,
　　E. Zimbabwe
HAB. Upland rain-forest and bamboo forest, especially in clearings and near edges; (1050–)1500–
　　2950 m.

SYN. *Urtica hypselodendron* A. Rich., Tent. Fl. Abyss. 2: 260 (1850)
　　? *Urera crenulata* Engl., V.E. 3(1): 53 (1915). Types: Zaire, Lake Kivu & Kenya, Mau, not precisely
　　indicated (B, syn.†)

NOTE. Well-defined by the characteristic deeply 4-lobed ♀ perianth, but sometimes in the literature,
　　e.g. Hauman in F.C.B. 1: 182 (1948), with respect to the material from "Forestier central" and the
　　illustration (repeated by Troupin, Fl. Rwanda 1: fig. 30/2A–C (1978)), confused with *U. flemingiana*
　　Lambinon.

Several collectors comment on the strong fibre in the stems, reported (*Glover, Gwynne & Samuel* 1496) to be used for string.

3. OBETIA

Gaudich., Voy. Monde Bonite, Bot., Atlas, t. 82 (1844); G.P. 3: 382 (1880); Engl. in E. & P. Pf. 3(1): 106 (1888); G.F.P. 2: 182 (1967); Friis in K.B. 38: 221–228 (1983)

Small to medium-sized shrubs or small trees, completely deciduous in dry season, dioecious. Stems herbaceous when young, wood soft and juicy, pith often wide; cystoliths dot-like or slightly elongated. Leaves alternate, usually clustered in terminal rosettes, petiolate, simple or lobed; stipules free, lateral. Inflorescences bracted panicles, in axils of fallen leaves or sometimes of current or young leaves. Male inflorescences much smaller than ♀ ones. Male flowers 5-merous, regular, with rudimentary ovary. Female flowers 4-merous, outer pair of tepals smaller than inner pair, both pairs enlarging and becoming thinly membranous in fruit; staminodes absent; ovary ovoid, oblique, with sessile penicillate stigma. Achene compressed, enclosed in the persistent, accrescent membranous perianth.

8 species in E. and S. Africa, Madagascar, the Mascarenes and Aldabra.

O. radula (*Bak.*) *B. D. Jackson* in I.K. 3: 323 (1894); Leandri in Fl. Madag. 56: 4, fig. 1 (1965); Troupin, Fl. Rwanda 1: 153 (1978); Friis in K.B. 38: 223 (1983). Type: Madagascar, Betsileo, *Langley-Kitching* (K, holo.!)

Sparsely branched tree, 5–13 m. high, presumably dioecious; trunk 0.2–0.5 m. in diameter at base, with main branching points either near the base or at the top, ultimate branchlets ± 0.8 cm. thick when dry, probably more when fresh, covered with large persistent stipules and stinging hairs; wood soft and with much juice in the thick bark and the wide pith; bark grey or brown, smooth to flaky, with very noticeable leaf-scars on the branches; the whole plant has an appearance not unlike a paw-paw, probably always completely deciduous for part of the year. Leaves always clustered at the end of the branches; stipules large, ovate to subcircular, 1.5–2.5 cm. long, 1.2–1.8 cm. wide, acute to acuminate, glabrous or ciliate, long-persistent, green but withering brownish; petiole 8–15 cm., pubescent and with many stinging hairs up to 2.5 mm. long; lamina ovate in outline, usually bullate, 15–25 cm. long and wide, palmately lobed or divided, rarely ± entire, with each lobe triangular, unlobed to ± clearly palmately lobed, base deeply cordate, margin (also of the lobes) serrate, apex (also of lobes) acuminate; lateral nerves 5–7 pairs, basal pair reaching the tip of the basal lobes; upper surface with short hairs and scattered stinging hairs, lower surface densely pubescent, and with scattered stinging hairs on the nerves; cystoliths punctiform, visible from above. Inflorescences in the axils of the current leaves, rarely persisting after the leaves have fallen, profusely branched panicles up to ± 20 cm. long, on peduncles up to ± 5 cm. long, pubescent and with numerous stinging hairs, flowers in small clusters at the branching points or scattered along the axes; ♂ and ♀ rather similar. Male flowers subsessile; perianth globular, up to ± 2 mm. in diameter, 5-merous, with rudimentary ovary present. Female flowers subsessile, 4-merous, perianth up to ± 0.8 mm. in diameter, compressed; stigma prominent penicillate, the whole flower yellow. Achene ± 1–1.5 mm. long. Fig. 3.

UGANDA. Ankole District: Mpororo Hill, Rwampara [Ruampara], Oct. 1932, *Eggeling* 1016!; Busoga/Mengo Districts: banks of the Nile, Busofu [perhaps error for Busoga], 1904, *Dawe* 88!; Mengo District: Kawanda Hill, Oct. 1937, *Chandler* 1981!
KENYA. Naivasha District: foot of the Kedong Escarpment, Oct. 1960, *Verdcourt & Lucas* 3002!; Kiambu District: Kikuyu Escarpment Forest, Oct. 1972, *O. J. Hansen* 745!; Masai District: Mt. Suswa, Oct. 1962, *Glover & Samuel* 3298!
TANZANIA. Maswa District: W. Serengeti, Moru Kopjes, Dec. 1956, *Greenway* 9157!; Mbulu District: Lake Manyara National Park, Marera River Gorge, Mar. 1964, *Greenway & Kanuri* 11371!; Morogoro, Nov. 1934, *E. A. Bruce* 139!
DISTR. U 2–4; K 1, 3, 4, 3/6, 6 7; T 1–3, 6; E. Zaire, Rwanda, Burundi, Madagascar
HAB. Local on rocky hill-sides in evergreen or semi-evergreen bushland (chiefly on hills of the basement complex, less commonly on basalt and lava), sometimes at rocky lake- and river-shores, and at margins of dry montane forest; 700–2000 m.

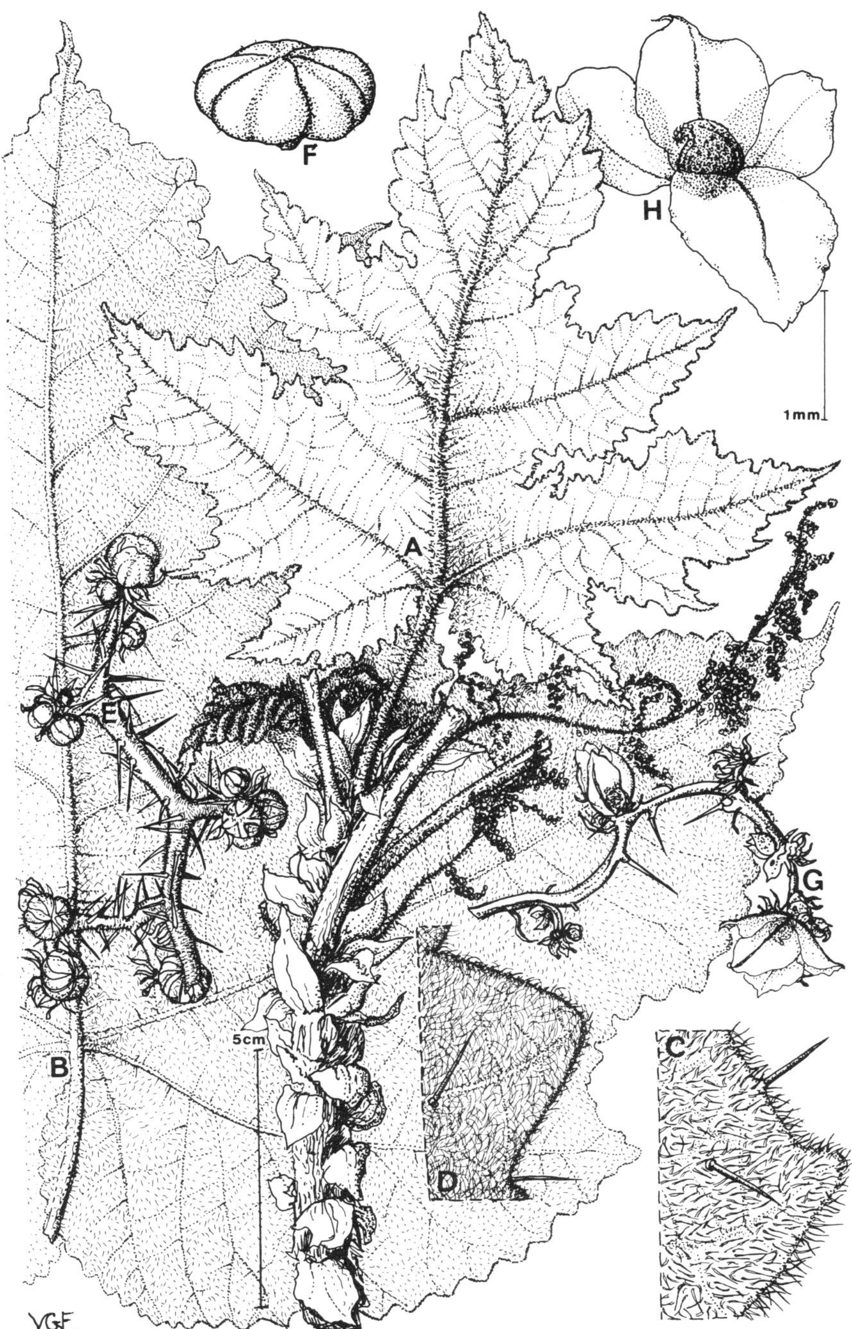

Fig. 3. *OBETIA RADULA* — **A**, flowering stem with young leaf; **B**, part of mature leaf; **C**, detail of leaf from above; **D**, detail of leaf from below; **E**, detail of male inflorescence; **F**, male flower in bud; **G**, detail of female inflorescence; **H**, female flower with accrescent perianth and young fruit. A, C & D, from *Chandler* 1981; B, from *Burtt* 2490; E, F, from *Richards & Arasululu* 26424; G, H, from *Evans* 777. Drawn by Victoria C. Friis.

SYN. *O. ficifolia* Gaudich. var. *heracleifolia* Pers., Syn. Pl. 2: 553 (1807). Type: unknown origin and
      collector, *Herb. Jussieu* 16852 (P-JUSS, lecto.!)
    Urera radula Bak. in J.L.S. 18: 279 (1881)
    Obetia morifolia Bak. in J.L.S. 20: 263 (1883). Types: Madagascar, without precise locality, *Bojer* &
      Imerina, *Baron* 1820 (both K, syn.!)
    O. pinnatifida Bak. in J.L.S. 20: 264 (1883); Rendle in F.T.A. 6(2): 244 (1917); T.S.K., ed. 2: 88
      (1936); Hauman in F.C.B. 1: 196 (1948); F.P.N.A. 1: 71 (1948); T.T.C.L.: 627 (1949); I.T.U., ed.2:
      441 (1952); K.T.S.: 579 (1961). Types: Madagascar, central part, *Baron* 1729 & *Baron* 1822 (both
      K, syn.!)
    O. laciniata Bak. in J.L.S. 20: 264 (1883). Types: Madagascar, central part, *Baron* 1721 &
      Andrangaloaka, *Parker* (both K, syn.!)

NOTE. The East African plants are undoubtedly conspecific with the material from Madagascar. In
South Africa the species is replaced by a pair of mutually closely related species, *O. tenax* (N. E. Br.)
Friis and *O. carrutersiana* (Hiern) Rendle.
    It is reported (*Edmonson* 33) that the roots are used by the Tugens in Kenya as a remedy against
barrenness; fibres from stem used for basket-making (*Kabuye* 82/53) and rope (*Glover et al.* 527).

4. GIRARDINIA

Gaudich. in Freyc., Voy. Monde, Bot.: 498 (1830); Endl., Gen. Pl.: 283 (1837); G.P. 3: 384
(1880); Engl. in E. & P. Pf. 3(1): 107 (1888); G.F.P. 2: 183 (1967); Friis in K.B. 36: 143–157
(1981)

Erect annual (or very short-lived perennial) herbs, monoecious or dioecious by
abortion, with long stinging hairs on all aerial parts; cystoliths punctiform. Leaves
alternate, petiolate, simple or mostly variously divided, margin coarsely serrate; stipules
intrapetiolar, fused almost to apex. Inflorescence unisexual, consisting of dense
elongated cymes in axils of upper leaves; the ♂ ones thinner and more spike-like than the
♀ ones. Male flowers pedicellate, 4–5-merous, regular, with large rudimentary ovary.
Female flowers sessile, with 3 almost completely fused tepals which ± enclose the ovary,
sometimes also with 1 minute, membranous, free tepal; staminodes absent; ovary ±
asymmetrical, reflexed, laterally compressed, with sessile, filiform stigma. Achene
laterally compressed, rugose, released from the perianth.

Two species distributed in the mountains of the Old World tropics from W. Africa to S. China and
Taiwan. The genus has the longest stinging hairs in the family, but apparently the sting is less severe
than that of some other genera, e.g. *Laportea.*

Tepals of ♂ flowers and largest tepal of ♀ flowers with horn-
    like appendage; leaves large, often up to 15 × 15 cm.,
    bullate, always with double serrate margin; mature stems
    more than 2 cm. in diameter . . . .   *1. G. bullosa*
Tepals without horn-like appendages; leaves smaller than 15 ×
    15 cm., not bullate, serrate (or rarely double serrate near
    the leaf-base); mature stems less than 2 cm. in
    diameter . . . .   *2. G. diversifolia*

1. G. bullosa (*Steud.*) *Wedd.* in Ann. Sci. Nat., sér. 4,1: 181 (1854); Rendle in F.T.A. 6(2):
265 (1917); Hauman in F.C.B. 1: 195 (1948); F.P.N.A. 1: 75 (1948); U.K.W.F.: 323 (1974);
Troupin, Fl. Rwanda 1: 155, fig. 31/1 (1978); Friis in K.B. 36: 145 (1981). Type: Ethiopia,
Tschenausa, *Schimper* 1409 (P [Herb. Steudel], holo.!, A, BM, G, L, LE, P, iso.!)

Annual or short-lived perennial monocarpic herb. Stems unbranched or little
branched near the top, somewhat lignified, hollow, up to 3 m. tall and 3.5 cm. in diameter;
bark greenish to dark brown, pilose and densely adorned with stinging hairs up to 8 mm.
long and 2 mm. thick. Leaves at anthesis mostly fallen from the lower part of the stem;
stipules broadly lanceolate, fused for ⅔ of their length, 1.8–2.4 cm. long, 0.5–0.8 cm. wide,
pubescent on the nerves and ciliate, otherwise glabrous; petioles 5–15 cm. long, pilose
and densely beset with stinging hairs up to 5 mm. long; lamina ovate, 18–35 cm. long,
16–27 cm. wide, base cordate, sometimes deeply, margin with 10–18 serrate, up to 2.5 cm.
wide teeth on each side, apex acute to acuminate; lateral nerves 5–7 pairs, basal pair
reaching 4th–6th tooth from apex; upper surface bullate, pilose, densely beset with
stinging hairs 2–4 mm. long, cystoliths punctiform, not clearly visible; lower surface

densely pubescent, especially on the nerves, and the larger nerves also with stinging hairs up to 5 mm. long. Inflorescences dense axillary panicles, but ♂ and ♀ very different; ♂ lax and little branched panicles, 10–20 cm. long with the flowers in clusters along the axes, each cluster up to 1 cm. in diameter; ♀ dense, repeatedly branched panicles with the ultimate branches forming small dichasia, the structure is at first 5–8 cm. long, but elongating during and after anthesis to form a dense network of branches with incurved tips, the whole inflorescence is ultimately up to 10 cm. long and 2.5 cm. wide, densely pubescent and beset with stinging hairs. Male flowers on pedicels 0.8–1.5 mm. long, articulate below the perianth; perianth 4-merous, ± 1.5 mm. in diameter, tepals with a subapical appendage ± 0.5 mm. long. Female flowers sessile, single in bifurcations of the dichasia, with 3 almost completely fused tepals and 1 short, free tepal, up to 1.5 mm. long, the middle of the fused tepal with a marked longitudinal keel; ovary enclosed in the perianth; stigma protruding, up to 3 mm. long, filiform. Achene ovoid, compressed, ± 2.5 mm. long, dark brown, usually verrucose.

UGANDA. Kigezi District: Sabinio Volcano, Dec. 1930, *B. D. Burtt* 2967!
KENYA. S. Nyeri District: Kiandongoro Forest, Aug. 1963, *Mathenge* 212!
TANZANIA. Masai District: Ngorongoro, Apr. 1941, *Bally* 2291!
DISTR. U 2;K 4;T 2; Sudan (Imatong Mts.), Ethiopia, E. Zaire, Rwanda, Burundi
HAB. Clearings and natural glades in upland rain-forest or at forest edges; 1800–2600 m.
SYN. *Urtica bullosa* Steud. in Flora 33: 259 (1850)

2. G. diversifolia (*Link*) *Friis* in K.B. 36: 145 (1981). Type: N. India, Sikkim,*J. D. Hooker* (K, neo.!, B, C, CAL, L, isoneo.!)

Erect, annual or short-lived perennial, monocarpic herb. Stems unbranched or little branched near the top, somewhat lignified, hollow, 1.5–2 m. tall, up to 1 cm. in diameter at base; bark often furrowed, greenish to dark brown, densely beset with stinging hairs 0.7–0.9 cm. long and with short stiff hairs. Leaves usually fallen from the lower part of the stem at anthesis; stipules linear-lanceolate, fused for at least ⅘ of their length, strigose to pubescent outside, glabrous inside, usually fallen at anthesis; petioles 3–15 cm. long, pubescent to pilose, densely beset with stinging hairs up to 0.6 cm. long; lamina ovate in outline, but varying from undivided to variously lobed or divided, 6–25 cm. long, 3.5–23 cm. wide, base cuneate, truncate, rarely subcordate (and usually only so in much divided leaves), margin in undivided leaves with 20–25 teeth on each side (in lobed or divided leaves with 10–20 teeth on the outer margin of the first lobes), apex (both of entire leaf and of lobes) acute or acuminate, sometimes with a terminal tooth 1–2 cm. long; lateral nerves 5–6 pairs, basal pair in undivided leaves reaching 10th–15th tooth from apex, in lobed or divided leaves ending in the tip of the basal lobes; upper surface with scattered stiff hairs and stinging hairs up to 7 mm. long, cystoliths punctiform, lower surface pubescent on the fine reticulation of nerves and with stinging hairs on the larger nerves. Inflorescence paniculate, but ♂ and ♀ very different; ♂ on peduncles up to ± 2 cm. long, overall up to 10 cm. long, little branched, with ♂ flowers in dense clusters which may be almost confluent along the axes; ♀ much branched panicles, with the ultimate branches forming dense dichasia, 2–3 cm. long at the time of anthesis, but elongating to 10–15 cm. during ripening of the fruit, densely beset with stinging hairs up to ± 5 mm. long. Male flowers on pedicels ± 1 mm. long; perianth ± 1 mm. in diameter, 5-merous; tepals without subapical appendage. Female flowers sessile in the bifurcations of the dichasia; perianth ± 1.5 mm. long, with 3 almost completely fused tepals, the 4th tepal usually suppressed; ovary enclosed in the perianth; stigma filiform, ± 2 mm. long, protruding. Achene ovoid to orbicular, compressed, ± 3.5 mm. long, smooth, brown. Fig. 4, p. 14.

UGANDA. Bunyoro District: Budongo Forest, Muhimbi Sample Plot, Sonso R., Nov. 1935, *Eggeling* 2281!; Kigezi District: Kachwekano,Jan. 1939,*Purseglove*533!; Mbale District: N. Elgon, Siti [Sit] R., Jan. 1953, *Dawkins* 776!
KENYA. Northern Frontier Province: Mt. Kulal, Oct. 1947, *Bally* 5554!; NE. Elgon, Oct. 1961, *Tweedie* 2240!; Masai District: Mt. Suswa, May 1963, *Glover* 3708!
TANZANIA. Maswa District: Moru Kopjes, Apr. 1962, *Greenway et al.* 10589!; Mbulu District: Mt. Hanang, May 1962, *Polhill & Paulo* 2293!; Songea District: Matengo Hills, Miyau, May 1956, *Milne-Redhead & Taylor* 10294!
DISTR. U 1–3; K 1, 3–6; T 1–8; widespread, from Senegal to S. Sudan (Imatong Mts.) and Ethiopia, S. to Angola, Zimbabwe, and South Africa (Transvaal); also on Madagascar, in Yemen, and Asia from Sri Lanka and India to S. China, Taiwan and Indonesia
HAB. Upland rain-forest, especially in clearings, natural glades and along edges, also in dry montane forest and in shady places in cultivations and in moist rocky places, caves and at base of rocky outcrops; 1100–2500 m.

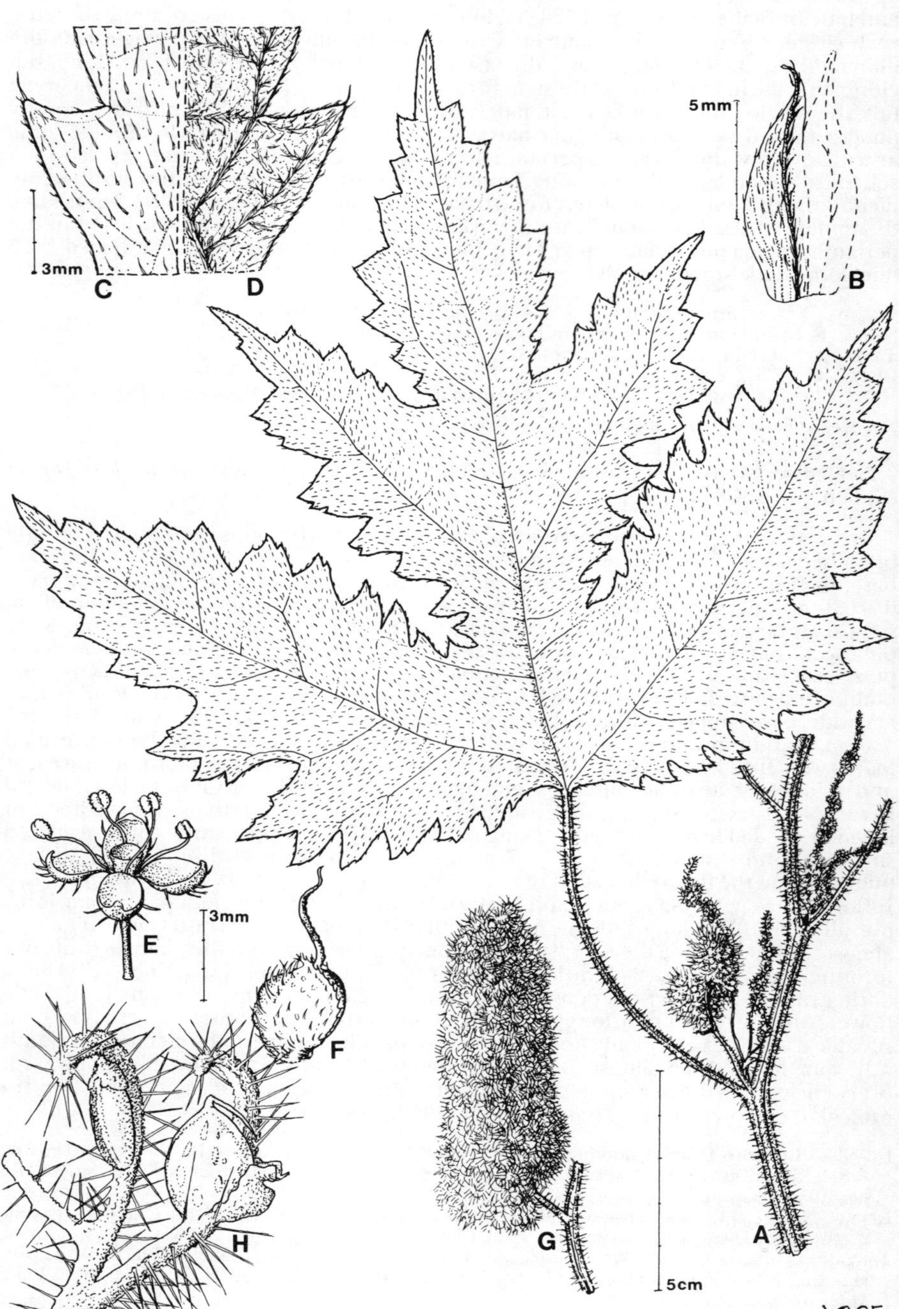

FIG. 4. *GIRARDINIA DIVERSIFOLIA* — **A**, flowering stem with leaf; **B**, fused pair of stipules; **C**, detail of leaf from above; **D**, detail of leaf from below; **E**, male flower; **F**, female flower; **G**, mature infructescence; **H**, detail of infructescence with two female flowers. A–D, G, from *Biegel* 2889; E, H, redrawn from Flore du Cameroun 8: fig 17/4, 6; F, redrawn from K.B. 36: 150, fig. 31. Drawn by Victoria C. Friis.

SYN. *Urtica diversifolia* Link, Enum. Pl. Hort. Berol. alt. 2: 385 (1822), *non* Blume (1825)
    U. heterophylla D. Don, Prodr. Fl. Nepal.: 59 (1825), *nom. illegit.*, *non* Vahl (1790). Type: Nepal, *Wallich* 4603 (K, holo.!)
    Girardinia heterophylla Decne. in Jacquemont, Voy. Ind. 4, Bot.: 151, t. 153 (1844); Letouzey, Fl. Cameroun 8: 110, t. 17 (1968); Wickens in K.B., Add. Ser. 5: 120 (1976); Troupin, Fl. Rwanda 1: 155, fig. 31/2 (1978). Based on *Urtica heterophylla* Vahl, Symb. Bot. 1: 76 (1790), *nom. superfl.*, including the type of *Urtica palmata* Forssk. (1775). Type: Yemen, *Forsskål* (C, syn.!) & Rheede, Hort. Ind. Malab. 2, t. 41 (1679)
    Urtica condensata Steud. in Flora 33: 260 (1850). Type: Ethiopia, Mt. Selleuda near Adua, *Schimper* 1888 (P [Herb. Steudel], holo.!, BM, BR, FI, G, K, L, LE, P, S, iso.!)
    Girardinia condensata (Steud.) Wedd. in Ann. Sci. Nat., sér. 4,1: 181 (1854); Rendle in F.T.A. 6(2): 266 (1917); Hauman in F.C.B. 1: 196 (1948); F.P.N.A. 1: 76 (1948); F.P.S. 2:278 (1952); F.W.T.A., ed. 2,1: 618 (1958)
    G. palmata Blume in Mus. Bot. Lugd.-Bat. 2: 158 (1856), *nom. illegit.*, *non Urtica palmata* Forssk. (1775). Type: India, Nilgiri Hills, *Leschenault* 54 (P, holo.!, also holotype of *G. leschenaultiana* Decne. (1844))
    G. vahlii Blume in Mus. Bot. Lugd.-Bat. 2: 158 (1856), *nom. superfl.* Based on *Urtica palmata* Forssk. (1775)

NOTE.   Leaf morphology in this species is extremely variable which has given rise to an extensive synonymy, only part of which is cited here; however, the leaves often vary very considerably on any individual and, moreover, the morphology of flowers and fruits does not support subdivision of this species. Not even a formal infraspecific taxonomy has been found feasible.
    The sting of this species is sharply felt, but lasts according to almost all reports from only a few minutes to a few hours. The long strong fibres of the stems are frequently used, e.g. (*Purseglove* 533) by the Bakiga, in Songea District (*Glegg* 32/48), in Sebei District (*Porter* 31), and in the Mbeya District (*Harewood* 84).

5. LAPORTEA

Gaudich. in Freyc., Voy. Monde, Bot.: 498 (1830); G.P. 3: 383 (1880); Engl. in E. & P. Pf. 3(1): 106 (1888); Chew in Gard. Bull., Singapore 21: 200 (1965); G.F.P. 2: 183 (1967); Chew in Gard. Bull., Singapore 25: 111 (1969), *nom. conserv.*

Fleurya Gaudich. in Freyc., Voy. Monde, Bot.: 497 (1830); G.P. 3: 382 (1880); Engl. in E. & P. Pf. 3(1): 106 (1888); G.F.P. 2: 183 (1967)

Annual or perennial herbs or shrubs, monoecious or dioecious. Stems with numerous stinging hairs. Leaves alternate, petiolate, simple, serrate; cystoliths linear or punctiform; stipules intrapetiolar, fused almost to apex. Inflorescences unisexual, occasionally with a few flowers of opposite sex intermixed, paniculate or appearing spike-like due to very reduced branches, sometimes with "interrupted spikes" due to some internodes being elongated. Male flowers usually pedicellate, 4–5-merous, with rudimentary ovary. Female flowers: pedicel sometimes winged, either laterally or on the dorsal and ventral side (as defined by the symmetry of the flower); tepals 4, unequal, lateral pair usually much larger than the median pair, median pair of unequal size, the dorsal one geniculate or not; staminodes absent; ovary asymmetrical, ovoid, laterally compressed; stigma sessile, filiform, or deeply trifid with filiform branches. Achene compressed, often with characteristic sculpturing on sides, consisting of a ring-like ridge enclosing a ± rugose surface, stipitate and shed separately (sect. *Laportea*) or sessile and shed with the persistent perianth (sect. *Fleurya*).

About 50 species, pantropical, extending into the temperate regions of N. America and E. Asia.
    The species of this genus are distinct, but sometimes quite difficult to identify due to the variation of leaf morphology and indumentum; a combination of these characters with floral characters offers the best chances for correct identification.

1. Stigma deeply trifid; at least the terminal part of the inflorescence appears as an interrupted spike because of reduction of lateral branches (lowermost branches sometimes developed) 2
Stigma simple; inflorescences branched 3

2. Plant annual, without stolons; inflorescence usually
      bisexual; stigmatic branches less than 1 mm. long;
      achene 1–1.5 mm. long; ♂ perianth with 4
      corniculate tepals . . . .   2. *L. interrupta*
   Plant perennial, usually with stolons; inflorescences
      unisexual, the ♂ ones erect, the ♀ ones prostrate or
      geocarpic; achenes longer than 3 mm.; stigmatic
      branches 2–3 mm. long; ♂ perianth with 5 non-
      corniculate tepals . . . .   3. *L. ovalifolia*
3. Female pedicels laterally winged (especially pronounced
      in fruit) . . . .   1. *L. alatipes*
   Female pedicel unwinged or dorsally-ventrally winged 4
4. Inflorescences with long thin glandular hairs . . . .   7. *L. aestuans*
   Inflorescences without such hairs 5
5. Leaves with large triangular teeth, up to ± 1 × 1 cm. on
      mature leaves . . . .   5. *L. mooreana*
   Leaves with inconspicuous teeth, never more than ± 0.3 ×
      0.3 cm. 6
6. Leaves lanceolate to elliptic, widest at the middle; stems
      erect, spreading or sprawling, not scrambling;
      persistent lateral tepals almost always enclosing the
      achene; ♂ flowers with flat, non-corniculate
      tepals . . . .   6. *L. lanceolata*
   Leaves ovate, widest below the middle; plants usually
      scrambling among other plants; persistent lateral
      tepals not covering the achene; ♂ flowers with
      corniculate tepals . . . .   4. *L. peduncularis*

1. L. alatipes *Hook.f.* in J.L.S. 7: 215 (1864); Rendle in F.T.A. 6(2): 252 (1917); Hauman
in F.C.B. 1: 194 (1948); F.P.N.A. 1: 71, t. 6 (1948); F.W.T.A., ed. 2,1: 620 (1958); Letouzey in
Fl. Cameroun 8: 117, t. 18 (1968); Chew in Gard. Bull., Singapore 25: 124 (1969); U.K.W.F.:
321 (1974). Type: Cameroon, Mt. Cameroon, *Mann* 1973 (K, holo.!)

Robust annual or short-lived perennial herb, up to 1.5 m. high, monoecious. Stems soft,
but usually somewhat lignified at base, erect or decumbent, little branched, usually with
sparse to dense stinging hairs when young, rarely glabrescent. Leaves usually fallen from
the lower part of the stem at anthesis; stipules lanceolate, united for more than ½ their
length, glabrescent, up to 1.5 cm. long; petiole (2.5–)5–12 cm. long, pubescent, usually
densely beset with stinging hairs, rarely completely without; lamina ovate, rarely elliptic-
ovate, (4–)8–15(–20) cm. long, (2–)4–8(–10) cm. wide, base broadly cuneate to subcordate,
margin on each side with 25–35 pointed teeth less than 0.5 cm. long, apex acuminate;
lateral nerves 4–6 pairs, basal pair reaching 10th–14th tooth from apex; upper surface
with dense or sparse scattered stinging hairs and punctiform cystoliths, lower surface
glabrous except for stinging hairs on the nerves, or pubescent. Inflorescences loosely
branching, unisexual panicles, with up to ± 8 side-branches bearing clusters of flowers; ♀
inflorescences in upper leaf-axils, up to 18 cm. long, on peduncles 4–9 cm. long, often
reaching above the leaves, with the individual flower-clusters of a characteristic fan-
shaped appearance, usually densely beset with stinging hairs; ♂ inflorescences in axils
below the ♀ ones, up to 10 cm. long, on peduncles 0.5–1 cm. long. Male flowers in clusters
2–3 mm. in diameter, on ± 1.5 mm. long pedicels, perianth 0.5–1 mm. in diameter,
4-merous, usually with stinging hairs. Female flowers: pedicel ± 1 mm. long, at anthesis
with narrow lateral ridges which broaden in fruit to wings up to 1.3 mm. wide, stinging
hairs dense to sparse, bristly hairs usually in 2 lines or sometimes evenly distributed;
tepals: the 2 lateral ones the largest, the dorsal one somewhat geniculate, becoming more
so after anthesis, when the fruit inflexes under the lateral wings of the pedicel; ovary with
filiform unbranched stigma up to 2 mm. long. Achene shortly stipitate, laterally
compressed, with centre of flattened sides irregularly rugose, pale brown, margin smooth
and black to brown, ± 2 mm. in diameter, dispersed without the perianth. Fig. 6/E, p. 22 &
7/E, p. 24.

UGANDA. Ruwenzori, Aug. 1938, *Purseglove* 313!; Kigezi District: Mt. Sabinio, June 1947, *Purseglove*
   2447! & Virunga Mts., Mgahinga, no date, *Eggeling* 1072!
KENYA. Embu District: Mt. Kenya, Castle Forest Station, Dec. 1972, *Gillett & Holttum* 20091!;
   Machakos/Masai Districts: Chyulu Hills, June 1938, *Bally* 7824!; N. Kavirondo District: Kakamega
   Forest, Oct. 1953, *Drummond & Hemsley* 4757!

TANZANIA.   Arusha National Park, Mt. Meru, crater bottom, Mar. 1971, *Richards* 26763!; Kilimanjaro, Bismarck Hill, Feb. 1934, *Greenway* 3853!; Iringa District: S. of Dabaga, Kibengu, Feb. 1962, *Polhill & Paulo* 1452!
DISTR.   U 2, 3; **K** 3–6; **T** 2–7; Cameroon, Ethiopia, E. Zaire, Malawi, Zambia, Zimbabwe, South Africa (E. Transvaal, SW. Natal, E. Cape Province)
HAB.   In undergrowth of upland rain-forest, *Hagenia* forest and mountain bamboo thicket, usually near water; 1400–3500 m.

SYN.   *Urticastrum alatipes* (Hook.f.) Kuntze, Rev. Gen. Pl. 2: 635 (1891)
   Fleurya urticoides Engl. in E.J. 33: 122 (1902). Types: Cameroon, Mt. Cameroon, Buea, *Preuss* 916 (B, syn., K, isosyn.!, BM, photo.!) & 1000 m., *Deistel* 388 (B, syn. †)
   Girardinia marginata Engl. in E.J. 33: 123 (1902). Type: Cameroon, Mt. Cameroon, Buea, *Preuss* 618 (B, holo. †)
   Fleurya urticoides Engl. var. *glabrata* Rendle in F.T.A. 6(2): 248 (1917). Type: Tanzania, Kilimanjaro, Marangu, *Volkens* 980 (BM, holo.!)
   F. alatipes (Hook.f.) N.E. Br. in Fl. Cap. 5(2): 547 (1925)

NOTE.   Fiercely stinging, the sting may reportedly last a day.
   Specimens of this species can be mistaken for *L. aestuans*, which is separable on the long soft glandular hairs, and for *L. peduncularis*, which is separable by the pentamerous ♂ flowers, or *L. interrupta*, which is distinguishable on the long bisexual inflorescence and the branched style.

2. L. interrupta (*L.*) *Chew* in Gard. Bull., Singapore 21: 200 (1965) & 25: 145 (1969); Letouzey in Fl. Cameroun 8: 125 (1968). Type: India, *Hermann* (BM-HERMANN, lecto.!)

Annual herb, up to 1 m. tall. Stems slightly woody at base, usually somewhat branched, puberulous to pubescent, with scattered stinging hairs up to 1.5 mm. long, sometimes raised on protuberances ± 1 mm. high; bark green to dark brown. Leaves crowded towards the top of the stem; stipules linear, connate ⅔ of their length or more, 3–5 mm. long, with short stiff hairs on the nerves; petiole (3–)5–11 cm. long, pilose to glabrescent, occasionally with a few stinging hairs; lamina ovate, 5–10 cm. long, 3–8 cm. wide, base broadly cuneate, truncate or subcordate, margin serrate, with 12–25(–30) teeth on each side, apex acute to acuminate; lateral nerves 3–6 pairs, basal pair reaching 6th–18th tooth from apex (about half way to the apex); upper surface with scattered stiff or stinging hairs, cystoliths elongated, lower surface with scattered stiff or stinging hairs on the nerves. Inflorescences usually bisexual in the axils of the upper leaves, or in the axils of fallen leaves just above soil-level, single, paniculate, but with very reduced side-branches so the inflorescences appear as interrupted spikes, of an overall length up to 30 cm., often with clusters of flowers from the base, but sometimes with a peduncle up to 10 cm. long, axis glabrous or puberulous, flowers in cymose clusters up to 1 cm. in diameter. Male flowers appearing first, on pedicels ± 1 mm. long; perianth ± 1 mm. in diameter, 4-merous, with corniculate tepals, sometimes with a few stinging hairs. Female flowers: pedicels ± 0.5 mm. long, unwinged or slightly dorsally and ventrally winged; tepals: the lateral ones largest, ± 0.8 mm. long, the dorsal one geniculate; ovary with a style of 3 filiform stigmatic branches, the longest up to 0.5 mm. Achene ± 1.75 mm. long, ovoid, laterally compressed, not stipitate, with a flattened wing-like structure along the edge, and a ridge surrounding the warted central part of each side, dispersed with the perianth. Fig. 6/A, p. 22 & 7/A, p. 24.

KENYA.   Northern Frontier Province/Lamu District: Boni Forest, Sept. 1961, *Gillespie* 406!; S. Nyeri District: Kahuroini, Oct. 1964, *Kibui* 53!
TANZANIA.   Kilosa District: Mikumi National Park, Kikoboga [Kikobola] R., June 1977, *Wingfield et al.* 4097!; Ulanga District: Kidatu–Sonjo, Mar. 1970, *Harris & Pócs* 4287!; Rungwe District: Kyimbila, no date, *Stolz* 727!
DISTR.   **K** 1/7, 4, 7; **T** 6–8; records of this species are widely and thinly scattered throughout the lowlands of tropical and South Africa, but many of these records are old; also widely distributed in tropical Asia and Oceania, N. to Japan, S. to Australia (Queensland), E. to Hawaii and Tahiti
HAB.   Lowland rain-forest, often along roads, riverine forest, moist places in wooded grassland; 0–550(–±1000) m.

SYN.   *Fleurya interrupta* (L.) Gaudich. in Freyc., Voy. Monde, Bot.: 497 (1830); Rendle in F.T.A. 6(2): 248 (1917); Hauman in F.C.B. 1: 190 (1948); U.K.W.F.: 321 (1974)
   Urtica lomatocarpa Steud. in Flora 33: 260 (1850). Type: Ethiopia, Tigre, in Mai-Mezano valley, *Schimper* 1471 (P [specimen marked "Herb. Steudel"], holo.!, A, BM, K, S, iso.!)

NOTE.   Recorded as fiercely stinging by collectors. Differs from all other African species of *Laportea* in the long interrupted spike-like inflorescences, but in many characters similar to *L. ovalifolia*, which differs in having stolons, unisexual inflorescences and much larger achenes. For distinction from the vegetatively similar *L. alatipes*, see under that species.

3. L. ovalifolia (*Schumach.*) *Chew* in Gard. Bull., Singapore 21: 201 (1965) & 25: 149 (1969); Letouzey in Fl. Cameroun 8: 131 (1968); Troupin, Fl. Rwanda 1: 157, fig. 31/3A–E (1978). Type: Ghana, without precise locality, *Thonning* (C, holo.!)

Perennial stoloniferous herb, sometimes almost entirely prostrate except for the erect ♂ inflorescences; erect shoots up to 1.25 m. high. Stems somewhat woody at base, pubescent to glabrescent, sometimes with short ± appressed stinging hairs. Leaves very variable in size and shape, usually restricted to the erect shoots, but sometimes also from the stolons; stipules linear–lanceolate, fused for at least ½ their length, up to ± 1 cm. long, with few stiff, perhaps stinging hairs, otherwise glabrous; petiole 5–10 cm. long, glabrous to puberulous, sometimes with a few appressed stinging hairs; lamina ovate, 3.5–10 cm. long, 2–8 cm. wide, base truncate to rounded or very broadly cuneate, margin serrate, on each side with 13–28 teeth less than 0.5 cm. long, apex acuminate, with a terminal tooth up 1 cm. long; lateral nerves 4–7 pairs, basal pair reaching 6th–12th tooth from apex (slightly more than half way to the apex); upper surface with short, stiff, presumably stinging hairs and punctiform cystoliths, lower surface with stiff, stinging hairs on the nerves. Inflorescences unisexual; ♂ inflorescences axillary or directly from the stolons, paniculate, with overall length up to ± 50 cm., usually with all the side-branches reduced to clusters, but occasionally with the lower branch developed, on a peduncle 10–20 cm. long, axes pubescent, somewhat juicy or fleshy, mostly conspicuously pinkish-brown, differing noticeably in colour and texture from the stems; ♀ inflorescences racemes or little branched panicle, up to 5 cm. long, sometimes in the axil of the upper leaves (not seen in material from Flora area), but mostly from stolons, and then usually geocarpic, axes thin, glabrous to puberulous, pinkish-red, up to ± 5 cm. long. Male flowers in cymose clusters up to 1.5 cm. in diameter, on pedicels up to 3 mm. long; perianth ± 2 mm. in diameter, 5-merous, tepals without appendages, sometimes with a few stinging hairs. Female flowers usually single or few together, on pedicels 0.5–2.5 mm. long; tepals subequal, up to ± 2 mm. long, the ventral one usually shorter; ovary slightly longer than the tepals; style with 3 branches, the longest up to ± 3 mm. long. Achene ovoid, laterally compressed, 3–4 mm. long, usually not stipitate and not reflexed, with a flattened wing-like outer zone, and a ridge surrounding a warted central area on each side, dispersed with the perianth. Fig. 6/C, p. 22 & 7/B, p. 24.

UGANDA. Mengo District: Mpanga Forest, near Mpigi, Jan. 1953, *Dawkins* 763!; Masaka District: Malabigambo Forest, SSW. of Katera, Oct. 1953, *Drummond & Hemsley* 4539!; Mengo District: Mabira Forest, E. of Kiwala, Apr. 1969, *Lye et al.* 2475!
KENYA. N. Kavirondo District: S. Kakamega Forest, near Kibiri Plot, May 1971, *Mabberley* 1115! & Kakamega Forest, Nov. 1969, *Bally* 13675!
TANZANIA. Kigoma District: W. of Lugufu, *Peter* 36516
DISTR. U 1–4; K 3, 5; T 1, 4; widespread in West and Central Africa, from Sierra Leone to S. Sudan, south to Angola and Zimbabwe
HAB. Lowland, transitional and upland rain-forest, swamp forest, along streams; 900–2000 m.

SYN. *Haynea ovalifolia* Schumach., Beskr. Guin. Pl.: 406 (1827)
    Fleurya podocarpa Wedd. in DC., Prodr. 16(1): 76 (1869); Rendle in F.T.A. 6(2): 251 (1917); Hauman in F.C.B. 1: 191, t. 20 (1948). Type: Nigeria, Angiana, *Barter* 92 (K, holo.!)
    F. podocarpa Wedd. var. *mannii* Wedd. in DC., Prodr. 16(1): 76 (1869); Rendle in F.T.A. 6(2): 252 (1917); Hauman in F.C.B. 1: 192 (1948). Type: Cameroon, Mt. Cameroon, *Mann* 1950 (K, holo.!)
    F. podocarpa Wedd. var. *amphicarpa* Engl., P.O.A. C: 163 (1895). Type: Tanzania, Bukoba, collector not specified, but according to Peter, F.D.O.-A. 2: 122 (1932), it must be *Stuhlmann* 1551 (B†)
    F. podocarpa Wedd. var. *fulminans* Hiern, Cat. Afr. Pl. Welw. 1(4): 989 (1900). Type: Angola, Golungo Alto, R. Mata de Quisucolo, *Welwitsch* 6265 (BM, holo.!)
    Pilea procumbens Peter, F.D.O.-A. 2: 126 & Appendix: 3, t. 12/1A,B (1932). Type: Tanzania, Kigoma District, Lugufu, *Peter* 36516 (B, holo.†)
    Fleurya podocarpa Wedd. var. *repens* [Hauman in F.C.B. 1: 192 (1948), *nomen.*] Lambinon in B.S.B.B. 91: 200 (1959). Type: Zaire, Forestier Centrale, Buta, *Robyns* 1209 (BR, holo.!)
    F. ovalifolia (Schumach.) Dandy in F.P.S. 2: 277 (1952); F.W.T.A., ed. 2,1: 619 (1958); U.K.W.F.: 321 (1974)

NOTE. The collection from Mt. Kenya (*Verdcourt* 2051), cited by Chew in Gard. Bull., Singapore 25: 152 (1969) as this species, is *L. alatipes*. I have not yet studied the collection from Mahenge (*Schlieben* 1814, BR) also cited as this species by Chew.
    The sting of this species is not reported as violent or painful.
    The species is easily recognized by its stolons, its characteristic ♂ inflorescences and its large achenes. Apparently a plant may sometimes consist of a rooted ♂ inflorescence only, without leaves, but the inflorescence probably always originates from a stolon attached to another plant.

The completely prostrate forms, that have been described under names such as *F. podocarpa* var. *repens* and *Pilea procumbens*, are here assumed to be habitat modifications, usually growing in very moist grassland, and have not been given formal taxonomic recognition.

4. L. peduncularis (*Wedd.*) *Chew* in Gard. Bull., Singapore 21: 201 (1965) & 25: 152 (1969); Friis in Bol. Soc. Brot., sér. 2, 58: 205 (1986). Lectotype chosen by Chew (1965): South Africa, Durban [Port Natal], at Yellowwood R., Kachu, *Drège* (G, lecto.!, K, iso.!)

Annual or short-lived perennial herb, monoecious, or dioecious by abortion. Stems soft, ± erect, decumbent or scrambling, sometimes rooting at lower nodes, up to 1.5 m. long, somewhat woody at the base, unbranched or branched from the base, pubescent to glabrescent, with few to fairly numerous stinging hairs ± 1 mm. long, sometimes raised on protuberances 0.5–1 mm. high; bark greenish to brownish. Leaves restricted to upper part of stem at anthesis; stipules linear-lanceolate, 4–10 mm. long, connate for about half their length, glabrous except for some stiff hairs on nerves; petiole 1–8(–11) cm. long, glabrous to puberulous, sometimes with stinging hairs (occasionally raised on short slender protuberances); lamina lanceolate to ovate, (2–)4–10 cm. long, (1–)2.5–7 cm. wide, base broadly cuneate to subcordate, margin serrate, on each side with (5(outside Flora area)–) 15–25 teeth shorter than 0.5 cm., apex acuminate, with apical tooth ± 0.8 mm. long; lateral nerves 5–8 pairs, basal pair reaching 6th–8th tooth from apex; upper surface with scattered, rather short stinging hairs, cystoliths punctiform, minute, lower surface with scattered stiff, probably stinging hairs on the nerves. Inflorescence unisexual, paniculate, single and usually restricted to the upper leaf-axils, sometimes reduced to an axis dichotomously branching once, with a few flowers on each branch, but mostly with many flowers in small cymose clusters; axes glabrescent or with a few stinging hairs; ♂ inflorescences in lower axils, on peduncles 1–3 cm. long, overall 3.5–9 cm. long, often with 2 or more glomerules of equal size terminating the branches; ♀ inflorescences in upper axils, usually repeatedly dichotomously branched, or with 1 branch overtopping the other, sessile or on peduncles, up to 2 cm. long, overall 3–6 cm. long. Male flowers in ± dense cymose clusters up to ± 1 cm. in diameter; pedicel 1–2 mm. long, with a narrow dorsiventral wing; perianth 1.5–2 mm. in diameter, (4–)5-merous; tepals with a dorsal, longitudinal crest and sometimes a few stinging hairs. Female flowers in ± dense cymose clusters up to 0.6 cm. in diameter; pedicel ± 0.5 mm. long, with a narrow dorsiventral wing; tepals: the lateral ones ± 0.8 mm. long, the dorsal one not geniculate, about as long as the lateral ones, the ventral one shorter; ovary with linear stigma, ± 1 mm. long. Achene ovoid, ± 1.5 mm. long, laterally compressed, almost sessile, somewhat reflexed when ripe, centre of flattened sides with a warted depression and smooth circular marginal ridge, dispersed with the perianth.

subsp. **peduncularis**

Erect, or, more often, prostrate or scrambling herbs, usually with an indumentum of stiff and stinging hairs mounted on thin protuberances; lamina ovate, base rounded to subcordate, margins on each side with 15–25 teeth. Fig. 6/B, p. 22 & 7/F, p. 24.

TANZANIA. Arusha District: E. of Mt. Meru, Ngurdoto Crater, June 1963, *Verdcourt* 3650!; Lushoto District: W. Usambara Mts., Mkuzi, Mkusu R., Mar. 1953, *Drummond & Hemsley* 1384!; Morogoro District: Uluguru Mts., Mgeta R., Hululu Falls, Mar. 1953, *Drummond & Hemsley* 1593!
DISTR. T 2,3,6; Malawi, Zimbabwe, Mozambique, South Africa (Transvaal, inland Natal, Cape Province), Swaziland
HAB. Transitional and upland rain-forest, especially in secondary growth and along paths and rivers; (300–)900–1650 m.
SYN. *Fleurya capensis* Wedd. var. *mitis* Wedd. in Archiv. Mus. Nat. Hist. Nat., Paris 9 (Monogr. Urtic.): 118 (1856). Types: South Africa, Cape Province, Galgebosch, *Drège* (P, syn.!, BM, K, isosyn.!) & Strandfontein and Matjiesfontein, *Drège* (P, syn.!, K, isosyn.!) & Doukamma, *Drège* (SAM, isosyn.!) & Natal, Yellowwood R., Kachu, *Drège* (P, syn.!, K, isosyn.!) & Van Staadens R., *Drège* (not traced)
[*F. capensis* sensu Wedd. in Archiv. Mus. Nat. Hist. Nat., Paris 9 (Monogr. Urtic.): 117, t. 1a, figs. 7,8 (1856); Rendle in F.T.A. 6(2): 249 (1917), *non* Wedd. (1854)]
F. peduncularis Wedd. in DC., Prodr. 16(1): 75 (1869)
F. peduncularis Wedd. var. *mitis* (Wedd.) Wedd. in DC., Prodr. 16(1): 76 (1869)
F. mitis (Wedd.) N.E. Br. in Fl. Cap. 5(2): 546 (1925)
Laportea caffra Chew in Gard. Bull., Singapore 25: 155 (1969). Type: South Africa, Cape Province, near George, *Schlechter* 2331 (Z, holo.!, A, BOL, BM, BR, GRA, K, P, PRC, PRE, S, SAM, STE, UPS, iso.!)
DISTR. (of species as a whole). Largely as for subsp. *peduncularis*, but in the coastal zone of

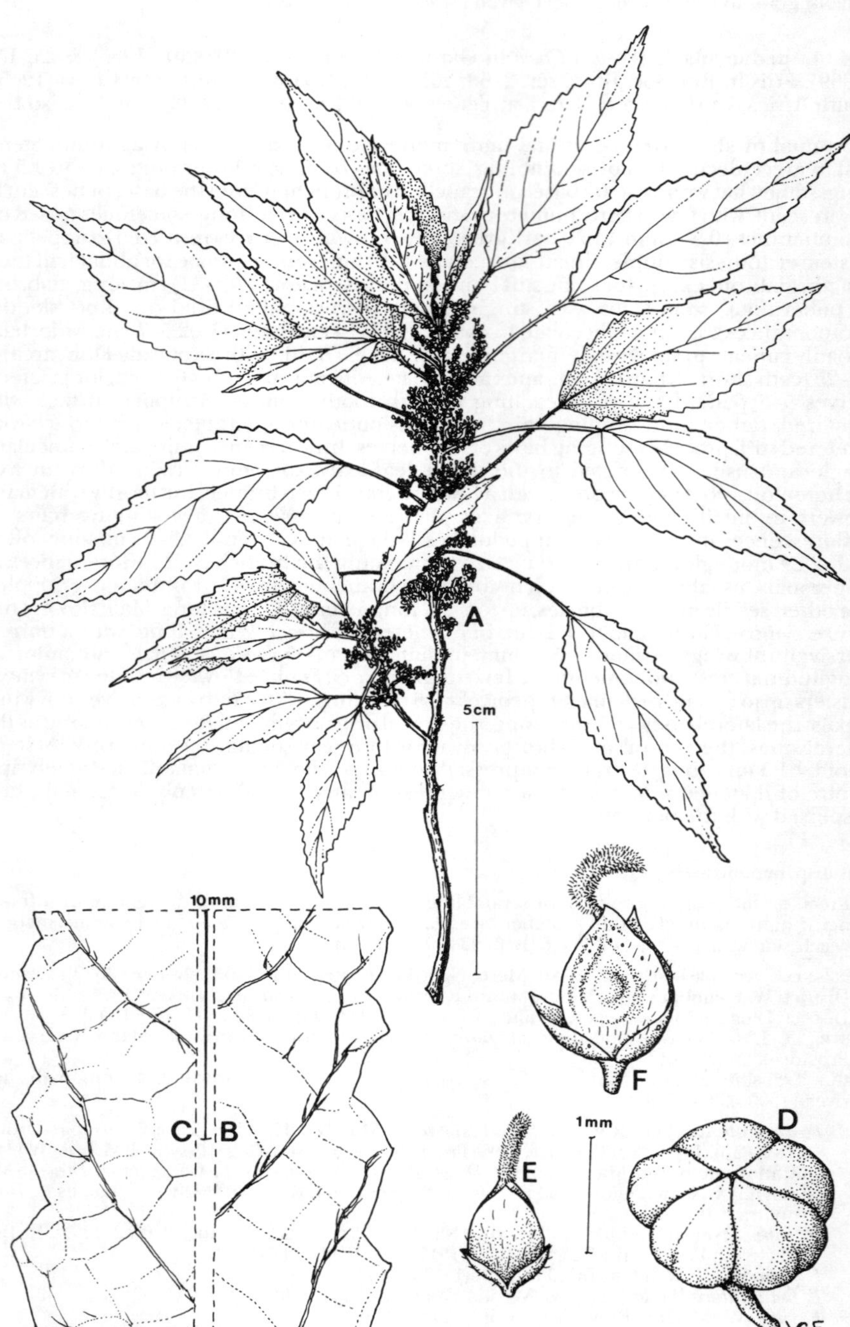

FIG. 5. *LAPORTEA LANCEOLATA* — **A**, flowering stem with leaves; **B**, detail of leaf from above; **C**, detail of leaf from below; **D**, male flower; **E**, female flower; **F**, old female flower with fruit. A–C, from *Adams* 27; D, E, redrawn from Chew in Garden's Bulletin Singapore 25, 1: 111–178 (1969). Drawn by Victoria C. Friis.

Mozambique and South Africa (coastal Natal) subsp. *peduncularis* is replaced by subsp. *latidens* Friis which has erect habit, glabrescent to glabrous stems, without stinging hairs, and leaves with cuneate leaf-base and with 5–8(–12) broad teeth on each side of lamina.

NOTE. *Fleurya capensis* Wedd. (1854) is based on *Urtica capensis* L.f. which is, according to the original material, *Didymodoxa capensis* (L.f.) Wedd., and the use of that named for the present species is therefore erroneous. In spite of *L. peduncularis* having almost consistently 5-merous ♂ flowers, the ♂ flowers in a few specimens, e.g. *Semsei* 3928, are 4-merous.

This species has reportedly only a moderately violent sting, perhaps due to the small number of stinging hairs.

5. L. mooreana (*Hiern*) *Chew* in Gard. Bull., Singapore 21: 201 (1965) & 25: 157 (1969); Letouzey in Fl. Cameroun 8: 127 (1968). Types: Angola, Mata de Pungo, *Welwitsch* 6256 & Pungo Andongo, *Welwitsch* 6276 (both BM, syn.!)

Annual (?and short-lived perennial) monoecious herb, up to ± 1(–?4) m. high, with few stems from the ± woody base. Stems puberulous to pubescent, frequently with stinging hairs 1–2 mm. long, raised on protuberances 1–2 mm. high; bark usually pale, greenish to brownish. Leaves crowded towards the top of the stems; stipules lanceolate, fused for ± half their length, ± 1 cm. long, with a few stiff hairs on the nerves; petiole 1.5–13(–15) cm. long, glabrescent, but distal part usually densely beset with raised stinging hairs as on stems; lamina broadly ovate to triangular, 4.5–15 cm. long, 3.5–13 cm. wide, base truncate to subcordate, margin coarsely serrate, on each side with 8–16 teeth, each tooth 0.6–1.3 cm. long and wide, with a distinct submarginal nerve on both sides, apex acuminate to caudate, terminal tooth up to ± 2 cm. long; lateral nerves 4–7 pairs, basal pair reaching 7th–8th tooth from apex; upper surface with scattered stinging hairs and punctiform cystoliths, lower surface glabrous to puberulous on the nerves, on which raised stinging hairs also frequently occur as on stem and petiole. Inflorescences unisexual or bisexual, paniculate, on peduncles 2–5 cm. long, up to ± 20 cm. long overall, single in the axils of the upper leaves; axes glabrous, frequently with raised stinging hairs, side-branches 10–20, up to ± 2 cm. long. Male flowers in separate inflorescences in the lower leaf-axils, or in the lower part of bisexual ones, on pedicels ± 1 mm. long; perianth 1–1.5 mm. in diameter, 4-merous, corniculate or with a subapical appendage up to 0.3 mm. long, sometimes with stinging hairs. Female flowers in inflorescences of the upper leaf-axils or in the upper parts of lower inflorescences, on pedicels ± 0.5 mm. long; tepals 4, or sometimes with 1 reduced, lateral ones ± 1 mm. long, the dorsal one slightly shorter, geniculate, usually with 1–few stinging hairs; ovary with linear stigma ± 0.5 mm. long. Achene ovoid, laterally compressed, 1–1.5 mm. long, sessile to slightly stipitate, on the flattened sides with a ridge surrounding a central warted depression, dispersed with the perianth. Fig. 6/D, p. 22 & 7/C, p. 24.

UGANDA. Karamoja District: S. of Kotido, Toror Hill, Sept. 1956, *Dyson-Hudson* 172!; Mengo District: Mabira Forest, Sept. 1967, *Tweedie* 3481! & 2 km. E. of Bujuku, Feb. 1969, *Lye* 1940!
TANZANIA. Ulanga District: Mahenge, Mar. 1932, *Schlieben* 1856!
DISTR. **U** 1, 2, 4; **T** 6; widely distributed in central tropical Africa, west to Nigeria, south to Angola, Zimbabwe and Mozambique
HAB. Lowland rain-forest, in the undergrowth, or on moist rocks in woodland or wooded grassland; 950–1250 m.

SYN. *Adicea mooreana* Hiern, Cat. Afr. Pl. Welw. 1(4): 991 (1900)
Pilea mooreana (Hiern) K. Schum. in Just's Bot. Jahresb. 28: 463 (1900)
Fleurya mooreana (Hiern) Rendle in F.T.A. 6(2): 250 (1917); F.W.T.A., ed. 2,1: 619 (1958)
F. funigera Mildbr. in N.B.G.B. 8: 278 (1923). Types: Cameroon, near Bamenda, Bali, *Conrau* 20 & Fumban, *Grewen* & same locality, *Versuchsanstalt Victoria* 292 & Joko, Jakong Village, *Thorbecke* 658 & NE. side of Mt. Njua, *Thorbecke* 715 (all B, syn.†)
F. urophylla Mildbr. in N.B.G.B. 8: 279 (1923). Type: NE. Zaire, Kivu, Muera, *Mildbraed* 2161 (B, holo.†)

NOTE. The syntypes of *Adicea mooreana* are all very weak plants with poorly developed inflorescences, hence Hiern's placement of the material in *Adicea* (a segregate of *Pilea*). However, there is no doubt that the material belongs to the taxon generally referred to as *L. mooreana*. One collection from the Flora area (*Dawkins* 652) was reportedly 4 m. high; elsewhere the species is reported to reach a height of ± 2 m. Collectors indicate that the sting of this species is painful.

6. L. lanceolata (*Engl.*) *Chew* in Gard. Bull., Singapore 21: 201 (1965) & 25: 161 (1969). Types: Tanzania, Usambara Mts., Mashewa, *Holst* 8833 (B, syn.†, K, isosyn.!, BM, photo.!) & Uluguru Mts., Mbora, *Stuhlmann* 9012 (B, syn.†) & Lukwangule Plateau, *Stuhlmann* 9223 (B, syn. †)

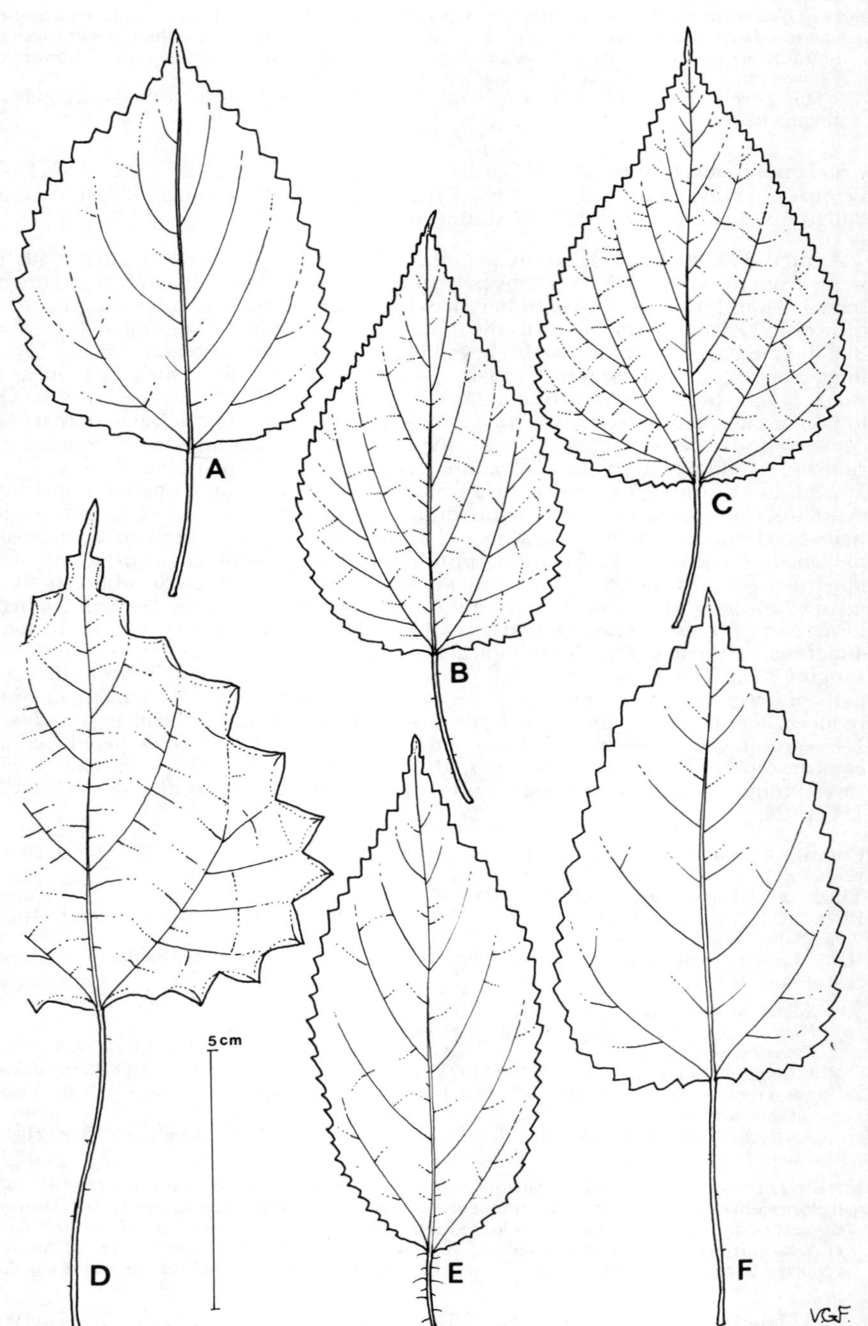

FIG. 6. *LAPORTEA SPP.* — Leaf-shape. **A**, *L. INTERRUPTA*; **B**, *L. PEDUNCULARIS* subsp. *PEDUNCULARIS*; **C**, *L. OVALIFOLIA*; **D**, *L. MOOREANA*; **E**, *L. ALATIPES*; **F**, *L. AESTUANS*. Drawn by Victoria C. Friis.

Presumably perennial and apparently dioecious, rarely evidently monoecious herb, sometimes shrubby with juicy branches, reportedly with spreading or sprawling habit, up to 1.5 m. high. Branches with yellowish-green bark, sometimes with large buds over the scars of fallen leaves, presumably for renewed growth; young branches glabrescent or glabrous, occasionally with stinging hairs up to 2 mm. long. Leaves crowded towards the end of the branches; stipules narrowly triangular, fused only at base, 2–3 mm. long, glabrous, mostly green; petiole 3–8 cm. long, glabrous except for a few occasional stinging hairs; lamina ovate to narrowly ovate or elliptic, 4–12 cm. long, 2.5–5 cm. wide, base cuneate to rounded, margin serrate, on each side with 15–20 broad teeth up to ± 7 mm. long, apex acute to acuminate; lateral nerves (3–)4–7 pairs, basal pair reaching 8th–10th tooth from apex; upper surface with scattered stinging hairs and minute punctiform cystoliths, lower surface with stinging hairs on the nerves and more elongated cystoliths. Inflorescences unisexual, rarely bisexual, paniculate, profusely branched from the base, peduncle indistinct, overall length of ♀ inflorescences 2–6 cm., ♂ ones 1.5–3 cm. long. Male flowers in clusters up to 0.8 cm. in diameter, on pedicels ± 2 mm. long; perianth ± 2 mm. in diameter, 5-merous, tepals not corniculate and without stinging hairs. Female flowers in small clusters on the branches; pedicels up to 0.4 mm. long, at first erect; tepals: the lateral pair up to 1 mm. long, the dorsal and ventral ones much smaller, subequal; ovary with filiform stigma ± 1 mm. long. Achene ovoid, laterally compressed, ± 1.5 mm. long, substipitate, almost completely covered by the accrescent lateral tepals, finally reflexed, on the sides with a ridge surrounding a small smooth area, dispersed with the perianth. Fig. 5.

KENYA. Kwale District: Marenje [Marenji] Forest, Sept. 1957, *Verdcourt* 1940!; Kilifi District: Chasimba, July 1969, *B.R. Adams* 27!
TANZANIA. Lushoto District: E. of Mashewa, July 1953, *Drummond & Hemsley* 3187!; Morogoro District: Nguru Mts., between Mhonda Mission and Manyangu Forest, Mar. 1953, *Drummond & Hemsley* 1852!; Uzaramo District: Pande Forest Reserve, May 1979, *Mwasumbi* 11736!
DISTR. **K** 7; **T** 3,6; ?E. Zaire
HAB. Lowland rain-forest, frequently sprawling over rocks in clearings or at forest edge, also on rocks in moist woodland; 50–900 m.

SYN. *Fleurya lanceolata* Engl., P.O.A. C: 163 (1895); Rendle in F.T.A. 6(2): 248 (1917); Hauman in F.C.B. 1: 190 (1948)

NOTE. The record from E. Zaire (Semliki District: *Humbert* 9001 (BR)) is cited both by Hauman in F.C.B. 1: 190 (1948) and Chew in Gard. Bull., Singapore 25: 163 (1969); it seems to represent an aberrant extension of the distribution of an otherwise strictly coastal species; the specimen should be further studied.
　　　Although the species has generally unisexual inflorescences, one specimen (*Renvoize* 1527) has clearly bisexual inflorescences.
　　　The sting is recorded as "extremely irritating" (*Faulkner* 4203).

7. L. aestuans (*L.*) *Chew* in Gard. Bull., Singapore 21: 200 (1965) & 25: 164 (1969); Letouzey in Fl. Cameroun 8: 121 (1968); Troupin, Fl. Rwanda 1: 157 (1978); Marais in Fl. Masc., 161, Urtic.: 5 (1985). Type: Probably grown in the Uppsala Botanical Gardens from seeds from (?) Surinam, *Linnaean Herbarium* No. 1111.14 (LINN, lecto.!)

Annual herb with somewhat fleshy stems, slightly woody at base, up to ± 1 m. tall or more, little branched, glabrescent to densely covered with stinging hairs up to ± 1 mm. long and soft glandular hairs 1–2 (–3) mm. long. Leaves restricted to the top of the stem; stipules linear-lanceolate, fused for ± half their length, glabrous except for a few stinging hairs on the midnerve; petioles 5–15 cm. long, usually densely covered with soft glandular hairs, with a few stinging hairs intermixed; lamina ovate to broadly ovate, 7–20 cm. long, 5.5–14 cm. wide, base broadly cuneate to cordate, margin serrate, on each side with 30–40 teeth up to 0.5 mm. long, apex acute or acuminate; lateral nerves 5–9 pairs, basal pair reaching 10th–12th tooth from apex; upper surface with scattered stiff and stinging hairs and punctiform cystoliths, lower surface with stinging hairs on the nerves. Inflorescences bisexual, rarely unisexual, profusely branched panicles on peduncles 4–10 cm. long, overall length 15–20 cm., in the axils of upper leaves and reaching above these, axes covered with glandular hairs, flowers in clusters ± 5 mm. in diameter. Male flowers scattered on the inflorescence and mixed with the ♀ ones or mostly at the base, soon falling, on pedicels ± 1 mm. long; perianth ± 1.5 mm. in diameter, 4–5-merous usually with glandular hairs. Female flowers densely clustered; pedicels ± 0.5 mm. long, usually slightly winged dorsiventrally; tepals: the two lateral ones ± 0.5 mm. long, the dorsal one half as long, slightly geniculate, usually with glandular hairs, the ventral one shorter;

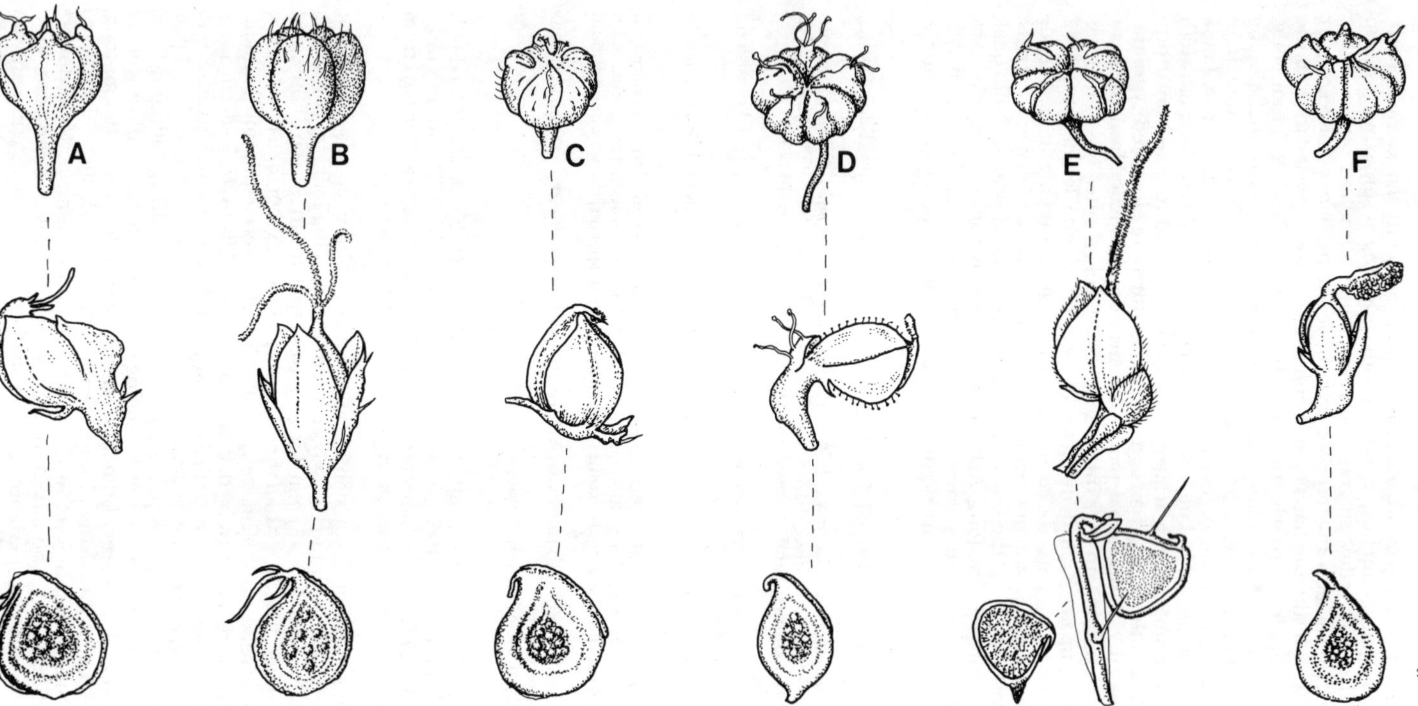

FIG. 7. *LAPORTEA SPP.* — Male flowers (top row), female flowers (middle row) and fruits (lower row) — **A**, *L. INTERRUPTA*; **B**, *L. OVALIFOLIA*; **C**, *L. MOOREANA*; **D**, *L. AESTUANS*; **E**, *L. ALATIPES*; **F**, *L. PEDUNCULARIS* subsp. *PEDUNCULARIS*; All redrawn approximately to scale from figures published by Chew in Garden's Bull. Singapore 25, 1: 111–178 (1969), except pedicel and fruiting perianth of **E** which is redrawn from Flore du Cameroun 8: Pl. 18, 7.

ovary with stigma ± 0.5 mm. long, filiform, ± recurved. Achene ovoid, stipitate, 1(–1.5) mm. long, laterally compressed, on each side with a ridge enclosing a warted area, finally reflexed, dispersed with the perianth. Fig. 6/F & 7/D.

UGANDA. W. Nile District: Arua, Dec. 1939, *Hazel* 417!; Mengo District: Kyagwe [Kiagwe], Bukasa Forest, Aug. 1932, *Eggeling* 888! & Kampala, close to shore of Lake Victoria, June 1938, *Chandler* 2431!
KENYA. Machakos District: Kibwezi [Kibwesi] Plains, Apr. 1938, *Bally* 7821! & Pungu R., Makueni, Mar. 1957, *Jarrett* MAK/5!; Teita District: Tsavo National Park East, E. of Voi, Jan. 1970, *Thulin* 293!
TANZANIA. Mbulu District: Lake Manyara National Park, Feb. 1964, *Greenway & Kanuri* 11192!; Ufipa District: N. Rukwa, Nziga Plain, Mar. 1959, *Richards* 12234!; Iringa District: Kilombero Scarp Forest Reserve, Gologolo Mt., Karenga, Mar. 1970, *Harris & Pócs* 4241!
DISTR. **U** 1, ?2, 4; **K** 4, 7; **T** 1–4, 6–8; **Z**; widely distributed in tropical Africa, west to Senegal, north to Ethiopia, south to Angola, Zimbabwe and Mozambique; also on Madagascar, in Yemen, and throughout tropical Asia from India to Indonesia; tropical America, West Indies
HAB. In disturbed places in forest or woodland, always in semi-shade, sometimes in rock-crevices; from near sea-level to 1300 m.

SYN. *Urtica aestuans* L., Sp. Pl., ed. 2: 1397 (1763)
    Fleurya aestuans (L.) Gaudich. in Freyc., Voy. Monde, Bot.: 497 (1830); Rendle in F.T.A. 6(2): 246 (1917); Hauman in F.C.B. 1: 188 (1948); F.W.T.A., ed. 2,1: 619 (1958); U.K.W.F.: 321 (1974)
    Urtica schimperiana Steud. in Flora 33: 259 (1850). Type: Ethiopia, Tigre, Modat, *Schimper* 1739 (P [Herb. Steudel], holo.!, K, iso.!)
    Fleurya aestuans (L.) Gaudich. var. *linnaeana* Wedd. in DC., Prodr. 16(1): 72 (1869), *nom. superfl.*, based on typical variety
    F. perrieri Leandri in Ann. Mus. Col. Marseille, sér. 6, vol. 7,8: 16 (1950), *non Laportea perrieri* Leandri (1950). Type: Madagascar, Montagne des Français, Camp d'Arbre, *Perrier de la Bâthie* 1753 & 16201 (both P, syn.!)
    Laportea bathiei Leandri in Fl. Madag. 56: 10 (1965). Type as for *Fleurya perrieri* Leandri

NOTE. The distribution of ♂ and ♀ flowers shows some variation in the Flora area. In *Jarrett* MAK/5 and a few other collections the upper inflorescences consist of ♀ flowers only, the lower ones, only up to ± 7 cm. long, are ♂. In *Eggeling* 888 there are exceptionally two inflorescences in a leaf-axil, a long predominantly ♀ one, and a shorter ♂ one. A few specimens (e.g. *Bally* 162 and *Pirozynski* 409) seem to have entirely ♂ flowers.

6. PILEA

Lindl., Coll. Bot, t. 4 (1821); G.P. 3: 384 (1880); Engl. in E. & P. Pf. 3(1): 108 (1888); G.F.P. 2: 185 (1967); Friis in K.B. 44(4) (1989), ined., *nom. conserv.*

Annual or perennial herbs, monoecious or rarely dioecious by abortion, sometimes with somewhat lignified basal parts or suffrutescent, mostly with juicy, translucent and turgescent stems, which may occasionally have scattered glandular cells in epidermis. Leaves opposite, petiolate, the two of a pair sometimes of unequal size; stipules intrapetiolar, fused completely to the apex, caducous or persistent; lamina entire, serrate or dentate, triplinerved; cystoliths linear, rarely punctiform. Inflorescences axillary or apparently terminal, unisexual or bisexual, pedunculate or sessile, bracted at least when young, in dichotomous cymes or in dense capitular clusters, either single or in branched paniculate inflorescences. Male inflorescences usually develop before or below the ♀. Male flowers usually 4-merous, rarely (2–)3-merous, often with tepals fused at base, mostly appendiculate (with a small, horn-like appendage at apex or subapical); rudimentary pistillode poorly developed or completely absent. Female flowers usually 3-merous, of which the tepals ± fused at base, the middle perianth-segment is usually the largest and ± appendiculate at apex, the lateral ones much smaller; a scale, representing a staminode, present inside each perianth-segment, hyaline, at first inflexed, reflexed at maturity and forcibly ejecting the achene; ovary erect, with apical sessile penicillate stigma. Achene compressed, ovate or elliptic; endosperm scarce. Embryo with large cotyledons.

A genus with at least 200 species in all tropical areas and most subtropical parts of the world, with the exception of Australia and New Zealand.

1. Margin of lamina entire; leaves (with petiole) less than 1 cm. long . . . . *10. P. microphylla*
    Margin of lamina always dentate or clearly crenate, at least in the terminal half; leaves (with petiole) more than 1.5 cm. long 2

2. Inflorescence apparently a terminal corymb with 4 equal or subequal parts, each supported by one of the leaves of the two uppermost (largest) pairs, the internodes of which are reduced; sometimes also smaller paniculate inflorescences in the axils of the lower leaves   *1. P. tetraphylla*

Inflorescence not as above, apparently terminal 3

3. Inflorescence always dense, sessile, axillary clusters with more than 10 flowers in the leaf-axils of the upper $\frac{2}{3}$–$\frac{1}{2}$ of the stem; peduncle never developed; ♀ flowers with 3 subequal perianth-segments, ♂ flowers consistently with 3 tepals, each with a subapical horn-like appendage more than 1 mm. long (note that *P. johnstonii* subsp. *kiwuensis*, with sessile or very shortly pedunculate ♀ inflorescences may key out here)   *3. P. rivularis*

Inflorescence not a sessile, axillary cluster, always with a distinct peduncle, at least in ♂ inflorescences; ♀ flowers with 3 unequal perianth-segments; ♂ flowers mostly with 2 or 4 tepals, not with subapical horn-like appendage or if present, then less than 1 mm. long 4

4. Inflorescence very short, consisting of small few-flowered cymose clusters along an apparently racemose axis up to 1.5(–2) cm. long, rarely with a few side-branches in the lower or middle part of the inflorescence, occasionally a single pedunculate cluster, up to 3.5 mm. in diameter, consisting mainly of ♂ flowers . . . *2. P. angolensis*

Inflorescence not as above, more than 2 cm. long at anthesis 5

5. Female inflorescences lax, thyrsoid or paniculate at anthesis, with the flowers single or in small cymes up to 0.3(–0.5) cm. in diameter 6

Female inflorescences at anthesis of one or more pedunculate corymbose clusters or heads, more than 0.3 cm. in diameter 8

6. Longest lamina of a plant more than 6 cm. long, always more than twice as long as broad; lateral nerves from midnerve (i.e. the nerves which connect the midnerve with the basal pair of lateral nerves running along the lamina near the margin) clearly individualised and unbranched, or only giving off 1–2 weak side-nerves in the basal half of lamina; ♀ inflorescences at anthesis (3–)4–10 cm. long, with up to 10 side-branches   *4. P. bambuseti*

Longest lamina of plant less than 6 cm., or, if longer, then lamina less than twice as long as wide; lateral nerves from midnerve not clearly individualised in the basal half of lamina, and dissolving into a network of almost equally developed fine nerves near the basal pair of lateral nerves, or at least giving off nerves of about the same thickness as the lateral nerves themselves; ♀ inflorescence at anthesis 1–4(–4.5) cm. long, usually with less than 5 side-branches 7

7. Teeth of leaf-margin with concave proximal edge and S-shaped or convex distal edge, almost always apiculate; lamina lanceolate or elliptic, rarely ovate, and if so then less than 7(–7.5) cm. long . . . . *5. P. sublucens*

Teeth of leaf-margin regular, triangularly pointed or acute, with convex sides, not apiculate; lamina ovate, rarely lanceolate, but if so longer than 7 cm. . . . . . *6. P. holstii*

8. Lamina clearly obovate or elliptic to circular, widest at or above the middle; stipules stiff, translucent, with rounded apex and cordate base, long persistent; plant usually drying dark grey or greyish-green . . . . *8. P. goetzei*

Lamina ovate or lanceolate, widest below the middle,
  sometimes circular (but then with deciduous stipules);
  stipules membranous, translucent, white or brown,
  triangular or broadly ovate to elliptic, deciduous or
  persistent; plant drying greenish or greenish-black 9
9. Stipules longer than (1.5–)3 mm., usually persistent, ovate
  to lanceolate, brown; lamina usually with spreading
  hairs on nerves beneath . . . .   *7. P. johnstonii*
  Stipules shorter than 1 mm., broadly triangular,
  translucent, rarely brown (do not confuse the stipules
  with the often very reduced or aborted axillary
  branches, which may be longer than the stipules);
  lamina glabrous beneath, rarely with a few spreading
  hairs on the nerves beneath near the base . . .   *9. P. usambarensis*

1. P. tetraphylla (*Steud.*) *Blume*, Mus. Bot. Lugd.-Bat. 2: 50 (1856); Rendle in F.T.A. 6(2):
270 (1917); Hauman in F.C.B. 1: 199 (1948); F.W.T.A., ed. 2,1: 621 (1958); Letouzey in Fl.
Cameroun 8: 173 (1968); U.K.W.F.: 323 (1974); Troupin, Fl. Rwanda 1: 165, fig. 33/3
(1978). Type: Ethiopia, near Adua, *Schimper* 74 (P [specimen marked "Herb. Steudel"],
holo.!, BR, K, P, iso.!)

Annual herb with fleshy pinkish stems, (3–)5–25(–40) cm. tall, unbranched or
branched from the lower 1–3 nodes, sometimes reaching flowering with the cotyledons
persistent; apparently monoecious. Stems glabrous, eglandular, with minute linear
cystoliths in epidermis. Leaves of a pair equal, but usually increasing in size from the
lower nodes to the top of the plant, while the length of nodes and petioles decreases, so
the uppermost 2 pairs appear to be verticillate and almost sessile; stipules thin,
membranous, persistent, ovate to subcircular, up to ± 4 mm. long and 5 mm. wide, with
subcordate base and rounded apex; petioles of lower leaves 0.4–2.5 cm. long, of upper,
subverticillate leaves 0.2–0.5 cm. long; lamina ovate, in the lower leaves 0.9–3 cm. long,
0.6–2.8 cm. wide, in the upper leaves up to 3.5 cm. long and 3 cm. wide, ovate, rarely
elliptic–subcircular, base cuneate to truncate, margin serrate, with (4–)5–12 teeth on each
side, apex blunt to acute, rarely acuminate, with a large terminal tooth; basal pair of
lateral nerves clearly marked, reaching 3rd–4th tooth from apex, giving off lateral nerves
to the lower teeth, other lateral nerves indistinct; upper surface with appressed hairs and
fine linear cystoliths, lower surface glabrous or with scattered hairs on the nerves, and
with scattered hydatodes. Inflorescences in the axils of the upper leaves (the largest
inflorescences corymb-like, forming a cross-shaped figure at the stem-apex, but
consisting of 4 inflorescences in the axils of the 4 uppermost leaves), paniculate, with
flowers in cymose glomerules; ♂ inflorescences or partial inflorescences apparently rare,
restricted to the axils of the upper leaves, where they occur together with purely ♀
inflorescences, with a structure as described above. Male flowers in small clusters on axes
up to 0.8 cm. long; pedicels 1–2 mm. long; perianth with 2 lobes and 2 stamens. Female
flowers: pedicel up to 1 mm. long; perianth (3–)4-merous, with 1 large, longitudinally
crested segment and (2–)3 very short ones, visible only as crenation on a short, basal tube;
4 hyaline, staminodal scales present, ± enclosing the ovary. Achene broadly ovate,
compressed, smooth or ± verrucose, brown, up to 0.8 mm. long. Fig. 8/A–E, p. 28.

UGANDA. Kigezi District: Bukimbiri, May 1950, *Purseglove* 3368! & Ishasha Gorge, Kanungu, June
  1952, *Lind* 38!; Mbale District: Bugisu [Bugishu], Buginyanya, Aug. 1932, A. S. *Thomas* 368!
KENYA. Elgeyo District: Kapchebelel, Oct. 1981, *Gilbert & Mesfin* 6467!; Nakuru/Masai Districts: Mau
  Forest Reserve, Aug. 1948, *Maas Geesteranus* 5735!; Kericho District: Sotik, Kibajet, Sept. 1949, *Bally*
  7443!
TANZANIA. Arusha District: Ngurdoto Crater, Oct. 1965, *Greenway & Kanuri* 12161!; Lushoto District:
  W. Usambara Mts., Matondwe Hill, May 1953, *Drummond & Hemsley* 2819!; Rungwe, Lusanje,
  Crater Lake, May 1975, *Hepper, Field & Mhoro* 5412
DISTR. U 2, 3; **K** ?1, 3–5; **T** 2, 3, 6–8; E. Nigeria, Cameroon, Bioko [Fernando Po], Sudan (Imatong
  Mts., Jebel Marra), Ethiopia, E. Zaire, Rwanda, Burundi, Malawi, NE. Zambia, also Madagascar
HAB. Upland rain-forest, along streams or in moist places on forest floor, rarely in moist, shaded
  places in woodland or open vegetation; weed in gardens; 1300–2800 m.

SYN. *Urtica tetraphylla* Steud. in Flora 33: 260 (Mar. 1850)
    Pilea quadrifolia A. Rich., Tent. Fl. Abyss. 2: 263 (late 1850). Types: Ethiopia, Tigre, Mt. Sholoda
      [Selleuda], *Schimper* 1680 (P, syn.!, K, isosyn.!) & *Quartin-Dillon* (P, syn.!)
    P. hypnopilea Bak. in J.B. 20: 267 (1882). Type: Madagascar, East Betsileo, *Baron* 117 (K, holo.!)
    P. modesta Bak. in J.L.S. 20: 265 (1883). Type: Madagascar, central part, *Baron* 907 (K, holo.!)

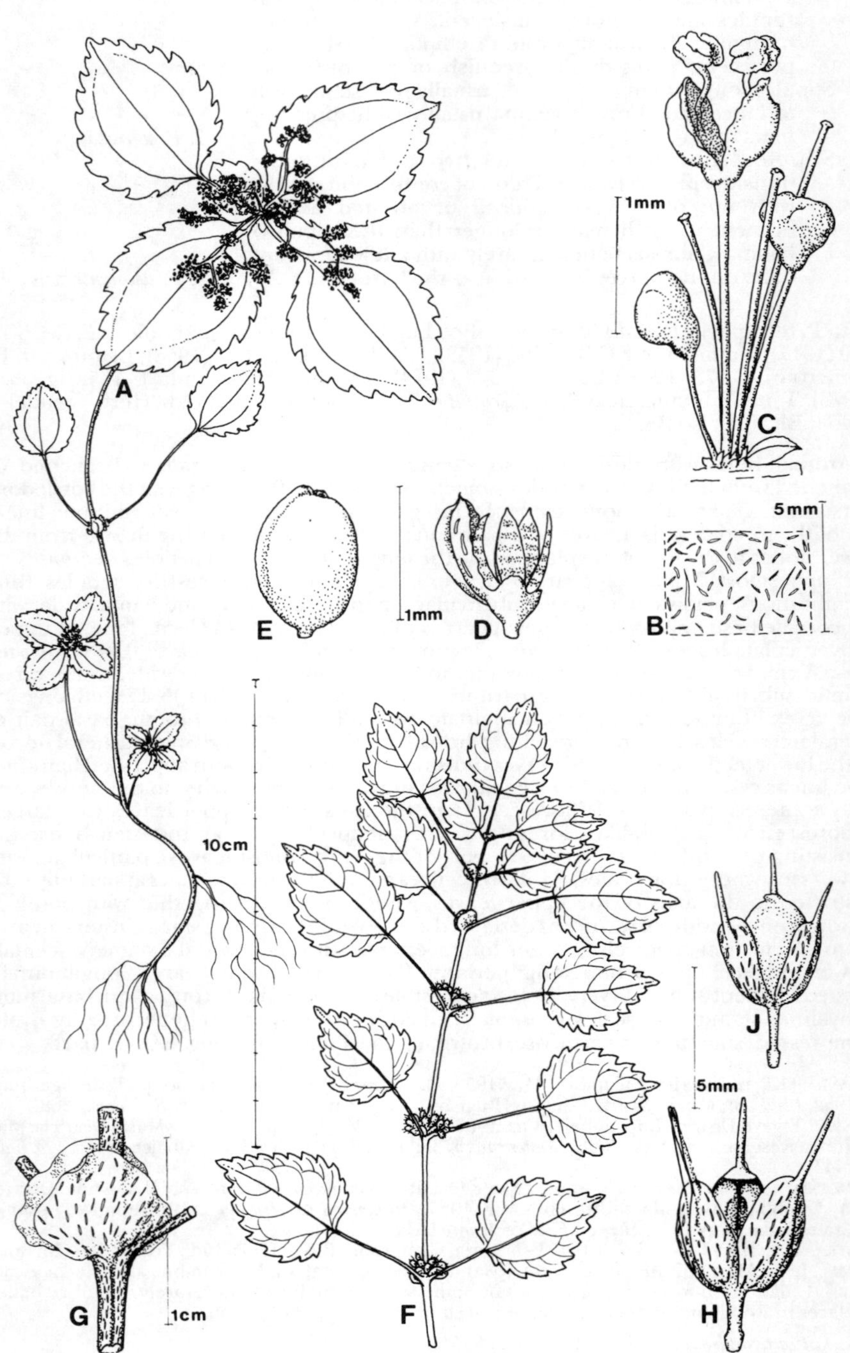

FIG. 8. *PILEA TETRAPHYLLA* — **A**, habit; **B**, detail of leaf from above; **C**, part of male inflorescence; **D**, female flower with achene ejected; **E**, achene. *P. RIVULARIS* — **F**, upper part of flowering stem; **G**, stipules; **H**, male flower; **J**, female flower. Drawn by Victoria C. Friis.

Adicea tetraphylla (Steud.) Kuntze, Rev. Gen. Pl. 2: 623 (1891)
Pilea tetraphylla (Steud.) Blume var. *major* Rendle in F.T.A. 6(2): 271 (1917). Type: Cameroon, Mt.
  Cameroon, Buea, *Preuss* 1001 (BM, holo.!)
P. tetraphylla (Steud.) Blume var. *hypnopilea* (Bak.) Leandri in Fl. Madag. 56: 46 (1965)

NOTE.   One collection (*Gillett* 13900) from Mt. Furolle was found among rocks and is the sole record
  from **K1**. The plants of this collection are dwarfed and not typical, and some doubt exists as to their
  identity. The specimens may possibly represent *P. angolensis* subsp. *angolensis*, which is another
  annual species well known from shade in woodland, typically on rocks.

2. P. angolensis (*Hiern*) *Rendle* in F.T.A. 6(2): 271 (1917); Hauman in F.C.B. 1: 200
(1948); F.W.T.A., ed. 2,1: 621 (1958); Letouzey in Fl. Cameroun 8: 160 (1968). Types:
Angola, Pungo Andongo, Barancos da Pedra Songue, *Welwitsch* 6258 & Petras de Guinga,
Welwitsch 6259 & without precise locality, *Welwitsch* 6272 (all BM, syn.!)

Annual erect herb, up to 20(–30) cm. high. Stems little branched, or with short
suppressed lateral shoots (consisting of a few nodes only) in the leaf-axils, glabrous, often
with orange or reddish glands in the epidermis. Leaves on the upper $\frac{1}{2}$–$\frac{2}{3}$ of the plant
usually with the largest lamina and the longest petioles; leaves of a pair equal or subequal;
stipules membranous, translucent and very inconspicuous, triangular, usually less than 1
mm. long; petiole 0.5–4 cm. long; lamina ovate to elliptic, (0.3–)1–3.5(–4.5) cm. long,
(0.2–)0.8–2(–3.5) cm. wide, base cuneate, rounded or subcordate, margin serrate, with
3–8(–13) teeth on each side (usually restricted to the apical $\frac{2}{3}$ of lamina), apex shortly
acuminate; lateral nerves 3–4 pairs, basal pair reaching 2nd–3rd tooth from leaf-apex;
both surfaces glabrous, upper surface with white dots between the numerous linear
cystoliths. Inflorescences usually 2 together in the upper leaf-axils, spike-like, with dense
cymose clusters along an axis up to 1.5 cm. long, rarely the axis branched to form a
paniculate inflorescence, or with only 1 cluster on a peduncle (occasionally 1
inflorescence and 1 reduced shoot with a few nodes in a leaf-axil). Male flowers either in
small, separate, cymose, globular inflorescences or in small clusters at the base of ♀
inflorescences; the ♂ flowers usually develop early, on pedicels ± 0.8 mm. long; perianth ±
0.8 mm. in diameter, 3–4-merous, each tepal with a subapical swelling. Female flowers
subsessile; perianth ± 0.8 mm. long, 3-merous, with 1 long, ovate perianth-segment with
an apical swelling, the 2 others much shorter, 3 hyaline staminodal scales present, shorter
than ovary. Achene ovate to lanceolate, compressed, 0.8–1 mm. long, pale brown.

subsp. **angolensis**

Generally as for the species, but leaves always circular to broadly ovate in outline, never more than
1.5 times as long as wide. Fig. 9/B, p. 32.

UGANDA.   Toro District: Bwamba, Buranga [Bulanga], Nov. 1935, *A.S. Thomas* 1528!; Mbale District:
  Bugisu [Bugishu], Nov. 1932, *Chandler* 1041!
TANZANIA.   Ulanga District: Kilombero Escarpment Forest Reserve, near Kidatu, Gologolo Mt.,
  Karenga, Mar. 1970, *Harris & Pócs* 4234!
DISTR.   **U** 2,3; **T** 6/7; in mountainous areas at medium altitudes throughout West Africa, from
  Guineé, Liberia and Sierra Leone to Sudan (Imatong Mts.) and Ethiopia, south to Angola
HAB.   In crevices of rocks or under boulders in forest or in moist woodlands, or on rocks along
  streams; 700–1500 m.

SYN.   *Adicea tetraphylla* (Steud.) Kuntze var. *angolensis* Hiern, Cat. Afr. Pl. Welw. 1(4): 990 (1900)
  Pilea divaricata Hauman in B.J.B.B. 17: 177 (1944). Type: Zaire, Kivu Province, Irumu, *Bequaert*
    2960 (BR, holo.!)

NOTE.   Known in East Africa only from the collections cited above. The other subspecies, subsp.
  christiaensenii (Lambinon) Friis, is characterized by constantly elongated leaves, always 2 times as
  long as wide or more; it is only recorded from the Walikale area of E. Zaire.

3. P. rivularis *Wedd.* in Archiv. Mus. Nat. Hist. Nat. 9 (Monogr. Urtic.): 266 (1856);
Letouzey in Fl. Cameroun 8: 163, t. 27 (1968). Type: Comoro Is., *Boivin* (P, holo.!)

Perennial herb with prostrate and ascending stems, up to 1 m. high; apparently mostly
dioecious, or at least each shoot unisexual. Stems little branched, often ± lignified at base,
glabrous, with linear cystoliths and occasionally reddish or orange glands in epidermis.
Leaves of a pair subequal or equal; stipules large and prominent, persistent, circular,
oblong or ovate, 6–10 mm. long, 4–8 mm. wide, transparent, with subcordate base and
rounded apex; petiole extremely variable, 0.4–5(–8.5) cm. long, glabrous or with a few
hairs at connection with lamina; lamina ovate to lanceolate, 0.5–8(–10.5) cm. long,

0.5–3.5(–6) cm. wide, base cuneate to truncate or subcordate, margin crenulate to serrate, with 4–15(–20) teeth on each side, apex acute to long acuminate; lateral nerves 3–4 pairs, basal pair reaching ± $\frac{2}{3}$ of the way to the apex before joining with the upper lateral nerves and arching towards the midnerve; both upper and lower surface with dense linear cystoliths, on the nerves of the lower surface often a ± dense indumentum of spreading hairs. Inflorescences sessile, dense glomerules, up to 1(–1.3) cm. in diameter, mostly unisexual, rarely with a few flowers of the opposite sex intermixed, single in the upper leaf-axils, with small linear bracts between the flowers. Male flowers 3-merous, on pedicels 1–2.5(–4) mm. long; perianth globular to ovoid, ± 2 mm. long (not including appendages); tepals equal, with prominent linear cystoliths in epidermis, each with a subapical filiform appendage 0.8–1.5 mm. long. Female flowers on pedicels 1–2 mm. long; perianth 1–1.5 mm. long, divided almost to the base, with 3 ovate, subequal lobes, the middle one with a subapical appendage up to 0.5 mm. long, lateral lobes $\frac{2}{3}$–$\frac{4}{5}$ as long as the middle lobe, usually also with a ± developed subapical appendage; staminodal scales about as long as perianth-segments. Achene brown to pale brown or whitish, ovoid, 1.5–2.5 mm. long, 1–1.5 mm. wide, ovoid, compressed, occasionally with a flat, marginal wing (rarely the wing dissolved into hooked bristles). Fig. 8/F–J.

UGANDA. Toro District: Ruwenzori, Kanyasabo, Jan. 1969, *Lye* 1339! & Bwamba Pass, Sept. 1932, *A. S. Thomas* 698!; Kigezi District: Ruhiza [Luhiza], May 1951, *Dawkins* 741!
KENYA. N. Nyeri District: Mt. Kenya, Naro Moru track, Dec. 1957, *Verdcourt* 2049!; Fort Hall District: Kimakia Forest Reserve, Feb. 1965, *Gillett & Kabuye* 16632!; Kericho District: Mau Forest Reserve, Aug. 1949, *Maas Geesteranus* 5605!
TANZANIA. Arusha National Park, Ngurdoto Crater, Oct. 1965, *Greenway & Kanuri* 11994!; Uluguru Mts., Lukwangule Plateau, Mar. 1953, *Drummond & Hemsley* 1539!; Songea District: Matengo Hills, Lupembe, May 1956, *Milne-Redhead & Taylor* 10518!
DISTR. U 1–3; K 3–7; T 2, 4, 6–8; E. Nigeria, Cameroon, Bioko [Fernando Po], E. Zaire, Rwanda, Burundi, Sudan (Imatong Mts.), Ethiopia, Malawi, Zimbabwe, South Africa (Transvaal); also on Madagascar and the Comoro Is.
HAB. Along streams and in other moist places in upland rain-forest or montane forest; 1250–3100 m.

SYN. *Pilea ceratomera* Wedd. in DC., Prodr. 16(1): 132 (1869); Rendle in F.T.A. 6(2): 269 (1917); Hauman in F.C.B. 1: 203 (1948); F.W.T.A., ed. 2,1: 621 (1958); U.K.W.F.: 323 (1974). Type: Bioko [Fernando Po], Clarence Peak, *Mann* 626 (K, holo.!)
    P. ceratomera Wedd. var. *mildbraedii* Engl. in Z.A.E. 1907–1908, 2: 191 (1911); Rendle in F.T.A. 6(2): 270 (1917); Peter, F.D.O.-A. 2: 128 (1932); Hauman in F.C.B. 1: 204 (1948). Types: Rwanda, Rugege, *Mildbraed* 1045 & Kisenyi [Kissenye], Ninagongo, *Mildbraed* 1355 (both B, syn. †)
    P. magambensis Engl., V. E. 3(1): 60, fig. 36G (1915); Peter, F.D.O.-A. 2: 129 (1932). Type: Tanzania, W. Usambara Mts., Magamba, Kwai, *Engler* 1298 & 1299 (both B, syn. †)
    P. worsdellii N.E. Br. in Fl. Cap. 5(2): 550 (1925). Types: South Africa, Transvaal, Houtbosch, *Schlechter* 4740 (K, syn.!) & Soutpansberg, *Worsdell* (K, syn.!)
    P. stipulata Hutch. & Dalz., F.W.T.A. 1: 443(1928) & in K.B. 1929: 19(1929). Type: Cameroon, 2500–2600 m., *Mann* 2011 (K, holo.!, P, iso.!)
    P. ceratomera Wedd. subsp. *glechomoides* Hauman in F.C.B. 1: 204 (1948), *nom. inval.*
    P. ceratomera Wedd. subsp. *glechomoides* Hauman forma *hypsophila* Hauman in F.C.B. 1: 204 (1948), *nom. inval.*
    P. rivularis Wedd. var. *stipulata* (Hutch. & Dalz.) Lambinon in B.J.B.B. 47: 255(1977)

NOTE. There is a considerable range of variation in leaf sizes and shapes in this species; however, the size generally diminishes with increasing altitude, with very small- and smooth-leaved forms in bamboo and *Hagenia* forest. It has not been possible to adapt a meaningful system of infraspecific classification to fit this variation. A fairly distinct high-altitude form ("var. *stipulata*") is small, with short internodes, comparatively large stipules, and small leaves. However, slightly larger but otherwise similar forms can be found locally at lower altitudes. Forms with comparatively narrow, lanceolate leaves have also been distinguished ("var. *mildbraedii*"), but all intermediates to the typical forms with ovate leaves exist.
    For the difficulties of distinguishing between *P. rivularis* and *P. johnstonii* Oliv. subsp. *kiwuensis* (Engl.) Friis in the small area where they are sympatric, see notes under the latter taxon.

4. P. bambuseti *Engl.* in Z.A.E. 1907–1908, 2: 190 (1911); Rendle in F.T.A. 6(2): 273 (1917); Hauman in F.C.B. 1: 199 (1948). Types: Rwanda, Kisenyi, Ninagongo, *Mildbraed* 1354 & Kisenyi, Bugoie [Bugoyer], *Mildbraed* 1476 & Rugege, *Mildbraed* 1044 (all B, syn. †)

Perennial herb with erect stems from prostrate base, up to 0.8(–1) m. high. Stems little branched, glabrous. Leaves of a pair subequal or one considerably smaller and with shorter petiole than the other; stipules persistent, broadly triangular, pointed, 0.8–2 mm. long; petiole 0.5–5.5 cm. long, glabrous or with scattered hairs in the distal part; lamina ovate to elongate or elliptic, 2.5–12.5 cm. long, 0.8–4.5 cm. wide, base usually ± oblique, subcordate to rounded, rarely cuneate, margin serrate, with 14–20 teeth on each side (more outside Flora area), each tooth usually regularly triangular, acute, not apiculate, leaf-apex long-acuminate, with the apical tooth 0.6–2 cm. long; basal pair of lateral nerves reaching apical tooth, lateral nerves (other than the basal pair) numerous, running straight from midnerve towards edge of lamina or giving off a few clearly weaker tertiary nerves, not disappearing into a network of tertiary nerves as in *P. holstii*; upper surface glabrous, with scattered, often inconspicuous linear cystoliths, lower surface with a few spreading hairs on the nerves and sometimes with hydatodes. Inflorescences flattened, diffuse panicles, on peduncles 1–2 cm. long, flowering part up to 10 cm. long and 5 cm. wide, with small cymose glomerules; ♂ inflorescences with 4–10 branches, glomerules 2–4 mm. in diameter; ♀ inflorescence with (3–)5–10 branches, glomerules ± 2 mm. in diameter. Male flowers on pedicels 0.4–0.8 mm. long; perianth ± 0.8 mm. in diameter, 4-merous, tepals without or with only weakly developed subapical appendage. Female flowers subsessile, 0.8 mm. long; middle perianth-lobe rounded to acute, without subapical appendage, the 2 lateral lobes only ½ as long as the middle lobe. Achene broadly ovoid, ± 0.8 mm. long, yellow to pale brown.

subsp. **bambuseti**

As described above; lamina lanceolate, margin on each side with up to 20 teeth. Fig. 9/K, p. 32.

UGANDA. Kigezi District: Impenetrable Forest, Mar. 1946, *Purseglove* 1964! & Rushasha, Dec. 1968, *Lye* 982! & Virunga Volcanoes, Mt. Sabinio, Nov. 1934, *G. Taylor* 2017a!

DISTR. U 2; E. Zaire, Rwanda, Burundi

HAB. In moist undergrowth of upland rain-forest; 1500–1950 m.

DISTR. (of species as a whole). As for subsp. *bambuseti*, with an extension represented by another subspecies, subsp. *aethiopica* Friis, in SW. Ethiopia

NOTE. Known in the Flora area only from the collections cited above; subsp. *aethiopica* differs in having ovate lamina, with 20–30 teeth on each side.

5. P. sublucens *Wedd.* in DC., Prodr. 16(1): 130 (1869); Rendle in F.T.A. 6(2): 273 (1917); F.W.T.A., ed.2, 1: 621 (1958); Letouzey in Fl. Cameroun 8: 171 (1968). Type: Bioko [Fernando Po], Clarence Peak, *Mann* 630 (K, holo.!)

Presumably perennial herb, up to 40 cm. high. Stems prostrate and ascending, little branched, fleshy, glabrous, with a few scattered reddish glands and linear cystoliths in epidermis. Leaves of a pair subequal to very unequal, sometimes one only ½ the size of the other, and with much shorter petiole; stipules broadly triangular, persistent, but inconspicuous, 6–8 mm. long; petiole 0.4–5 cm. long, glabrous, or with a few hairs arranged in tufts at connection with lamina; lamina ovate to elliptic, 1.2–8(–9) cm. long, 0.6–3.5(–4.7) cm. wide, base rounded, margin serrate, with 7–12 teeth on each side, each tooth with a convex or S-curved distal edge and a concave proximal side, apex of tooth usually markedly apiculate, leaf-apex acute or usually acuminate; lateral nerves numerous, basal pair of lateral nerves reaching the level of 2nd or 3rd tooth from apex before arching inwards and joining the upper lateral nerves, rarely the basal pair reach the apical tooth of lamina, lateral nerves above the basal pair dissolve in the lower ⅔ of lamina in a dense network of tertiary nerves; upper surface glabrous, with scattered, linear cystoliths, lower surface glabrous with scattered hydatodes. Inflorescences unisexual, ± lax panicles, 0.6–6 cm. long, with few, alternate branches and flowers in small, ± diffusely cymose clusters; ♂ inflorescences: peduncle 0.8–1.3 cm. long, clusters usually few, up to 6 mm. in diameter; ♀ inflorescences more branched than the ♂ ones, clusters 2–3 mm. in diameter. Male flowers on pedicels 0.6–1 mm. long; perianth ± 1 mm. in diameter, 4-merous, tepals with a subapical appendage. Female flowers on pedicels ± 0.5 mm. long; perianth ± 1 mm. long, the middle lobe the longest, without subapical appendage, lateral lobes less than ½ as long. Achene long, ovoid, laterally compressed, 1–1.5 mm. brown. Fig. 9/F, p. 32.

UGANDA. Kigezi District: Impenetrable Forest, Apr. 1948, *Purseglove* 2682! & Ishasha Gorge, Nov. 1946, *Purseglove* 2279!

DISTR. U 2; Liberia, Ivory Coast, Nigeria, Bioko [Fernando Po], Cameroon, Gabon, Zaire, Rwanda

HAB. In rain-forest at medium altitudes; 1250–1550 m.

SYN. *P. preussii* Engl. in E.J. 33: 123 (1902). Types: Cameroon, Mt. Cameroon, N. of Buea, *Preuss* 573 & W. of Buea, *Preuss* 953 (both B, syn.†, K, isosyn.!)
  P. chevalieri Schnell in Rev. Gén. Bot. 57: 280, fig. 2 (1950); Letouzey in Fl. Cameroun 8: 167 (1968). Type: Guineé/Liberia/Ivory Coast, Mt. Nimba, *Schnell* 952 (P, syn.!) & Nuon, *A. Chevalier* 21117 (P, syn.!, K, isosyn.!)

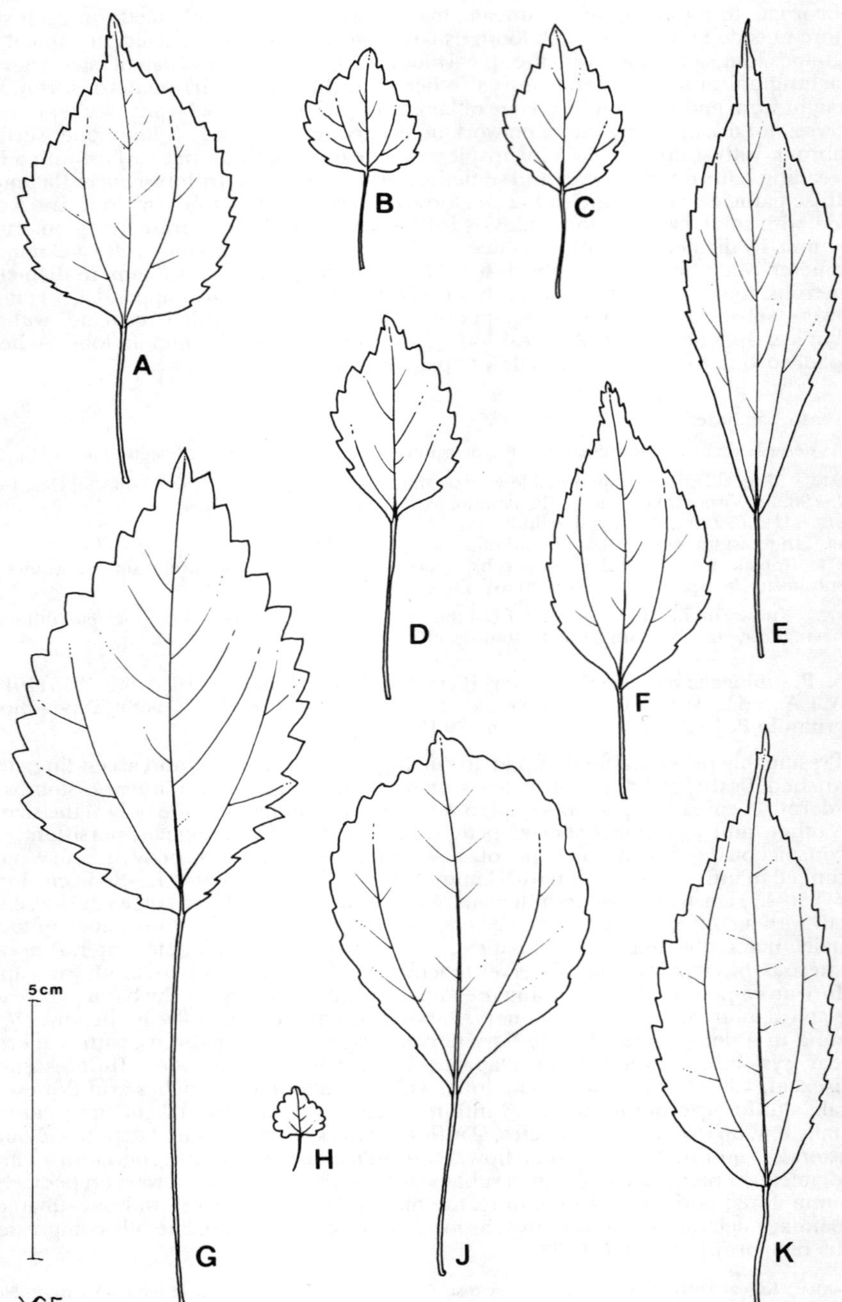

FIG. 9. *PILEA SPP.* — shape of leaves. **A**, *P. HOLSTII*; **B**, *P. ANGOLENSIS* subsp. *ANGOLENSIS*; **C**, *P. JOHNSTONII* subsp. *RWANDENSIS*; **D**, *P. USAMBARENSIS* var. *USAMBARENSIS*; **E**, *P. USAMBARENSIS* var. *ENGLERI*; **F**, *P. SUBLUCENS*; **G**, *P. JOHNSTONII* subsp. *JOHNSTONII*; **H**, *P. USAMBARENSIS* var. *VERONICIFOLIA*; **J**, *P. GOETZEI*; **K**, *P. BAMBUSETI* subsp. *BAMBUSETI*. A, from *Greenway* 1032; B, from *Adames* 378; C, from *Stauffer* 1029; D, from *Verdcourt* 207; E, from *Drummond & Hemsley* 1290; F, from *Adames* 580; G, from *Polhill & Paulo* 1628; H, from *Greenway & Brenan* 8300; J, from *Thulin & Mhoro* 3187; K, from *Lye* 982. Drawn by Victoria C. Friis.

6. P. holstii *Engl.*, P.O.A. C: 163 (1895). Type: Tanzania, E. Usambara Mts., Derema [Nderema], *Holst* 2245 (B, syn. †, K, isosyn.!) & W. Usambara Mts., Kwa Mshusa [Msuza], Nkongoi [Kongoi], *Holst* 9077 (B, syn. †, P, Z, isosyn.!)

Erect perennial herb with creeping basal part, up to 1 m. high or more. Erect stems unbranched or little branched, glabrous. Leaves of a pair subequal or one with much longer petiole than the other; stipules broadly triangular, persistent but inconspicuous, up to 0.8 mm. long; petiole 0.4–6.5(–8) cm. long, glabrous; lamina ovate to lanceolate, sometimes with oblique basal part, 1.2–8(–12) cm. long, 0.8–4.5(–7) cm. wide, with rounded or cuneate base, margin with 5–15(– 18) teeth, each tooth with straight or slightly convex sides, acute, never markedly apiculate, leaf-apex long-acuminate, with an apical tooth up to 1.5(– 2) cm. long; lateral nerves numerous, basal pair reaching the apical tooth or the 1st or 2nd tooth from apex before joining the upper lateral nerves, other lateral nerves in the lower ⅔ of lamina branch and dissolve into a network of fine nerves towards the edge of lamina; upper surface glabrous, with regularly scattered linear cystoliths, lower surface glabrous with hydatodes. Inflorescences much branched, panicles with the flowers in small, diffuse cymose clusters; ♂ inflorescences on peduncles 1–2 cm. long, flowering part 1.5–4.5 cm. long, with 2–4 branches, clusters up to 7 mm. in diameter; ♀ inflorescences more lax than the ♂ ones, on 1–2 cm. long peduncles, flowering part 2–5.5 cm. long, with 3–6 branches, clusters ± 2 mm. in diameter. Male flowers on pedicels ± 1 mm. long; perianth ± 1 mm. in diameter, 4-merous, tepals with subapicular appendage. Female flowers subsessile; perianth ± 0.8 mm. long, largest perianth-lobe lanceolate, acute, without appendage, the 2 lateral lobes only ½ as long. Achene ovoid, compressed, 0.8–1 mm. long, sometimes with a flattened margin, brown. Fig. 9/A.

KENYA. Kwale District: Jombo Hill, Jan. 1978, *Gilbert et al.* 4963!; Kilifi District: Chasimba, Sept. 1974, *B.R. Adams* 85!
TANZANIA. W. Usambara Mts., Kwashemshi [Kwamshemishi]–Sakare road, Aug. 1953, *Drummond & Hemsley* 3170! & E. Usambara Mts., Amani, July 1929, *Greenway* 1660!; Morogoro, Nov. 1934, *E. M. Bruce* 111!
DISTR. **K** 7; **T** 3, 6; E. Zaire
HAB. In lowland rain-forest; 200–1400 m.

NOTE. The collection *Drummond & Hemsley* 1725 from the Uluguru Mts. has unusually large leaves with oblique, cuneate base, thus approaching some forms of *P. bambuseti*; however, the characteristics from the leaf nervation place the collection in *P. holstii*. One collection from the E. Usambara Mts. (*Richards* 15349, from Bulwa [Burwa] Tea Estate) is said to come from an altitude of 1800 m.; however, the highest points round Bulwa reach only about 1000 m.

7. P. johnstonii *Oliv.* in Trans. Linn. Soc., Bot., ser. 2, 2: 349 (1887); Rendle in F.T.A. 6(2): 273 (1917); Hauman in F.C.B. 1: 200 (1948); U.K.W.F.: 323 (1974). Type: Tanzania, Kilimanjaro, *Johnston* (K, holo.!, BM, iso.!)

Perennial herb with prostrate and ascending stems, up to 0.6 m. high. Stems juicy, glabrous. Leaves of a pair subequal; stipules ovate or triangular, prominent, brown, membranous, (1.5–)3–7.5(–8.5) mm. long, 0.8–4 mm. wide, base subcordate, apex rounded; petioles 0.3–8(–10) cm. long, glabrous or with a few scattered spreading hairs near the connection with lamina; lamina broadly ovate to ovate, rarely elliptic or lanceolate, (1–)1.5–8.5(–12.5) cm. long, (0.8–)1–7.5(–8.5) cm. wide, base cuneate, truncate or subcordate, margin serrate, with 6–24 teeth on each side, apex short-acuminate; lateral nerves 4–10, basal pair reaching 4th–7th(–10th) tooth from the apex; glabrous on both surfaces or with spreading hairs on the nerves on the lower surface, upper surface with regularly arranged linear cystoliths, lower surface sometimes with clearly marked hydatodes. Inflorescences usually 2 together in the upper leaf-axils, always consisting of a ± pedunculate corymbose head, but sometimes with 2 side-branches below the head, rarely with repeated further branching under each head and up to 7 heads on each peduncle. Male inflorescences (heads) 5–12 mm. in diameter, on peduncles 0.8–6 cm. long; ♀ inflorescences (heads) subsessile or on peduncles 1–5 cm. long, 4–10 mm. in diameter. Male flowers on pedicels 0.8–2 mm. long; perianth ± 1 mm. in diameter, 4-merous; tepals with up to 0.5 mm. long subapical appendages. Female flowers subsessile; perianth 0.8(–1) mm. long, middle lobes the largest, with subapical appendage. Achene ovoid 0.8–1 mm. long, often with a ± marked flattened edge.

KEY TO INFRASPECIFIC VARIANTS

1. Peduncle of ♀ inflorescence at anthesis absent or hardly
     longer than the corymbose head of flowers; largest
     lamina less than 2.5(–3) cm. wide, lanceolate . . .   b. subsp. **kiwuensis**
   Peduncle of ♀ inflorescence at anthesis much longer than
     the corymbose of flowers; largest lamina wider than
     (2.5–)3 cm., ovate to lanceolate 2
2. Peduncle of ♀ inflorescence with repeatedly branched axes,
     almost always with more than 2 corymbose heads of
     flowers; head usually with a basal involucre of brown
     membranous bracts at base; stipules more than 3 mm.
     wide, usually long persisting . . . .   a. subsp. **johnstonii**
   Peduncle of ♀ inflorescence with only 1 corymbose head of
     flowers (rarely with 2 heads, but then one above the
     other on an apparently racemose axis), cymose heads
     usually without or with only very few and insignificant
     brown bracts at the base; stipules less than 3 mm. wide,
     often fallen from the lower leaves . . . . .   c. subsp. **rwandensis**

a. subsp. **johnstonii**

   Habit as described above. Stems up to 0.6 m. high, usually rather juicy. Leaves: stipules ovate,
prominent, persistent, brown, membranous, 4–7.5(–8.5) mm. long, 3–4 mm. wide, base subcordate,
apex rounded; petiole 1.2–8(–10) cm. long, often with a few hairs near the connection with lamina;
lamina broadly ovate or ovate, rarely elliptic or broadly lanceolate, (1–)1.5–8.5(–12.5) cm. long,
(1–)1.5–7.5(–8.5) cm. wide, base subcordate or cordate, rarely broadly cuneate, margin serrate, with
10–24 teeth on each side, apex broadly acuminate; basal pair of lateral nerves reaching 5th–7th(–
10th) tooth from apex; glabrous on both sides, rarely with a few scattered hairs above and on the
nerves beneath. Inflorescences pedunculate heads, sometimes repeatedly branched, with 2
branches from below each head and producing up to 7 heads on each peduncle; ♂ inflorescences
8–12 mm. in diameter, on peduncles 1–4.5 cm. long; ♀ inflorescences 4–8 mm. in diameter, on
peduncles 1–3.5 cm. long. Male flowers as above. Female flowers and achenes as above. Fig. 9/G, p.
32.

UGANDA.   Toro District: Ruwenzori, Kivata, 1893–94, *Scott Elliot* 7742! & Nyamwamba [Namwamba]
     valley, Jan. 1935, *G. Taylor* 3065!; Mbale District: Elgon, Mt. Nkokonjeru, Dec. 1924, *Snowden* 945!
KENYA.   Northern Frontier Province: Marsabit Forest, Aug. 1968, *Faden* 68/634!; N. Nyeri District: Mt.
     Kenya, Naromoru, Dec. 1957, *Verdcourt* 2052!; Masai District: Shabal Taragwa, Ol'Pusimoru
     Sawmill, May 1961, *Glover, Gwynne & Samuel* 1416!
TANZANIA.   Kilimanjaro, Bismarck Hill, Feb. 1934, *Greenway* 3855!; W. Usambara Mts., Mkuzi, Apr.
     1953, *Drummond & Hemsley* 2056!
DISTR.   U 2, 3; K 1, 3–7; T 2, 3, 7; Zaire (Ruwenzori), Sudan (Imatong Mts.), S. Ethiopia, E. Zimbabwe
HAB.   In upland rain-forest and montane forest, sometimes as an epiphyte; 1450–2900(–3600) m.

SYN.   *P. johnstonii* Oliv. var. *runssorensis* Engl., P.O.A. C: 163 (1895). Type: Zaire/Rwanda/Uganda, Ruwenzori,
     Stuhlmann 2730 (B, holo. †); Zaire, Ruwenzori, Parc National Albert, *Germain* 1101 (BR, neo.!)

NOTE.   The extention of the upper limit of altitudinal range from 2900 to 3600 m. is based on only
     one collection (*Mearns* 1280, from Mt. Kenya). The collection *G. Taylor* 3065 approaches subsp.
     rwandensis in general habit, but has stipules of typical subsp. *johnstonii*, and is, therefore, included
     here.

   b. subsp. **kiwuensis** (*Engl.*) *Friis* in K.B. 43: 648 (1988) Type: Zaire, Kivu, Virunga Volcanoes, Mt.
Sabinio [Sabinyo], *Mildbraed* 1692 (B, holo.†); Virunga Volcanoes, Sabinio [Sabinyo], *Bamps* 3243 (K,
neo.!, BR, isoneo.!)

   Habit as for typical subsp., up to 0.8(–1.5) m. high. Stems glabrous or with scattered hairs on upper
part. Leaves: stipules (2.5–)3–6.5 mm. long, 2–3 mm. wide, base subcordate to cuneate, apex rounded,
brown, membranous, persistent; petiole 0.3–5 cm. long; lamina lanceolate, 2.2–6 cm. long, 0.8–2.5(–
3.5) cm. wide, base cuneate, margin serrate, with 10–14 teeth on each side, apex acuminate; basal pair
of lateral nerves reaching 4th–5th tooth from leaf-apex; upper surface with scattered stiff hairs, lower
surface with spreading hairs on the nerves. Inflorescences sessile to shortly pedunculate heads; ♂
inflorescences subsessile or on peduncles up to 0.8(–1.7) cm. long; ♀ inflorescences sessile or on
peduncles up to 0.5 cm. long. Male flowers on pedicels up to 2 mm. long; perianth ± 1 mm. in
diameter, 4-merous, tepals with subapical appendages. Female flowers on pedicels ± 1 mm. long;
perianth ± 1 mm. long, otherwise as in typical subsp. Achene ± 1 mm. long, sometimes with flattened
margin.

UGANDA.   Kigezi District: Virunga Volcanoes, saddle between Muhavura and Mgahinga, Nov. 1954,
     Stauffer 738! & E. slope of Mgahinga, Nov. 1954, *Stauffer* 828!

DISTR. U 2; E. Zaire, Rwanda
HAB. In open *Hagenia* forest or woodland; 3000–3250 m.

SYN. *P. kiwuensis* Engl. in Z.A.E. 1907–1908, 2: 190 (1911); Rendle in F.T.A. 6(2): 275 (1917); Hauman
    in F.C.B. 1: 202 (1948)
    P. elatostematifolia Hauman in B.J.B.B. 17: 178 (1944) & in F.C.B. 1: 204 (1948). Type: Zaire, Kivu,
    Karisimbi, *Humbert* 8575 (BR, holo.!)

NOTE. Known from the Flora area only from the two collections cited above, of which *Stauffer* 738 is
    sterile. The subspecies would appear to be a high altitude form of subsp. *johnstonii*, and occurs
    typically on the Zaire side of the Virunga Volcanoes in *Erica arborea* bushland.
      Specimens of this taxon with completely sessile inflorescences are very similar to forms of *P.
rivularis* with lanceolate leaves, which also occur in the same general area. However, the following
characters may be useful in separating the two taxa: *P. johnstonii* subsp. *kiwuensis* has usually
entirely cuneate leaf-bases, while the narrow-leaved forms of *P. rivularis* are ± subcordate at the
very base of lamina; the ♂ flowers also offer good characters, in *P. johnstonii* subsp. *kiwuensis* the ♂
flowers are 4-merous and the appendages are up to ± 0.5 mm. long, while in *P. rivularis* the ♂
flowers are 3-merous and the apical appendages usually ± 1 mm. long; the ♀ flowers of *P. rivularis*
can be recognized by the lateral perianth-segments which are ⅔–⅘ as long as the middle lobe,
while in *P. johnstonii* subsp. *kiwuensis* the lateral lobes are much shorter than the middle lobe; the
achene in *P. johnstonii* subsp. *kiwuensis* is more narrow (2 times as long as wide or more) than the
achene of *P. rivularis*.

      c. subsp. **rwandensis** *Friis* in K.B. 43: 648 (1988). Type: Zaire, Kivu, Kahusi-saddle, between Bukavu
and Walikale, *Stauffer* 1092 (Z, holo.!, BR, iso.!)

Habit of plant more slender than in typical subsp., up to ± 0.6 m. high. Stems almost completely
glabrous. Leaves: stipules (1.5–)2–3.5 mm. long, 0.8–1.2 mm. wide, triangular, persistent but more
slender than in the 2 previous subsp. and usually only clearly visible at the upper 2–5 nodes; petiole
(1–)1.5–4.5(–6) cm. long, glabrous or almost so; lamina ovate to elliptic, (1.5–)3–8.5 cm. long,
(1.2–)1.4–4 cm. wide, base broadly cuneate, margin serrate, with 6–9 teeth on each side, apex
acuminate; basal pair of lateral nerves reaching 4th or 5th tooth from apex; upper surface with
scattered hairs, lower surface with few to many spreading hairs on nerves, rarely glabrous, hydatodes
often present. Inflorescences small cymose pedunculate heads, usually with one head on each
peduncle (the axis very rarely branching below the first head, so the inflorescence consists of 3
heads); ♂ inflorescences 0.5–0.8 cm. in diameter, on peduncles 2.5–6 cm. long; ♀ inflorescences
0.5–1 cm. in diameter, on peduncles 1–5 cm. long. Male flowers on pedicels 0.8–1.5 mm. long; tepals
with subapical appendages. Female flowers on pedicels ± 1 mm. long. Achene ± 1 mm. long, with a ±
flattened margin. Fig. 9/C, p. 32.

UGANDA. Ankole District: Kalinzu Forest, Kyamahungu [Kyamahunga], Jan. 1953, *Dawkins* 770!;
    Kigezi District: Impenetrable Forest, Rushasha, Dec. 1968, *Lye* 980! & Ishasha Gorge, Apr. 1946,
    Purseglove 2030!
TANZANIA. Mbeya District: Poroto Mts., Kikondo Camp, Jan. 1961, *Richards* 13956!; Rungwe District:
    Kyimbila highlands, Apr. 1912, *Stolz* 1197! & Ngozi [Wentzel-Heckmann] Crater, Mar. 1932, *St.
Clair-Thompson* 674!
DISTR. U 2; **T** 7; E. Zaire, Rwanda, Burundi, SW. Ethiopia
HAB. Upland rain-forest, dry montane forest; 1250–1950 m.

NOTE. Typical material of this subspecies exists in large quantities from E. Zaire, Rwanda and
    Burundi, especially from altitudes between 1800 and 2500 m. From the Flora area it is known only
    from the collections cited above. Of these *Lye* 980 and *Purseglove* 2030 are typical, while the
    collections *Richards* 13956, *Stolz* 1197 and *St. Clair-Thompson* 674 somewhat approach slender
    forms of subsp. *johnstonii*; however, they belong here on account of the technical character of the
    stipules. The last collection cited, *Dawkins* 770, belongs here only on account of the stipule
    characters.

      8. P. goetzei *Engl.* in E.J. 28: 379 (1900); Rendle in F.T.A. 6(2): 274 (1917). Type:
Tanzania, Uluguru Mts., Lukwangule Plateau, 2400 m., *Goetze* 286 (B, holo.†); Uluguru
Mts., Lukwangule Plateau, 2000 m., *Drummond & Hemsley* 1641 (K, neo.!, BR!, EA, isoneo.)

Erect or ascending perennial herb with conspicuous horizontal rhizome. Stems juicy,
often quadrangular, up to 80 cm. high, but sometimes ± entirely prostrate, glabrous, or
with a few scattered hairs. Leaves of a pair subequal; stipules ovate, persistent, stiff and
usually not hyaline, 5–9 mm. long, 2–5 mm. wide, with cordate base and rounded to blunt
apex; petiole 1–6 cm. long, glabrescent; lamina circular (rarely ovate) to obovate, 1.1–9.5
cm. long, 0.7–6.3 cm. wide, base truncate to cuneate, margin in the apical ½–⅔ crenate to
serrate, apex acute to broadly acute; lateral nerves (3–)5–9 pairs, basal pair reaching
2nd–3rd(–4th) tooth from apex, other lateral nerves little branched; upper and lower
surface glabrous and of the same colour, upper surface with small, densely and regularly
placed linear cystoliths in epidermis, lower surface with scattered hydatodes in epidermis.

Male inflorescences single on glabrous peduncles 1–8 cm. long, flowers in cymose heads, 0.5–1.3 cm. in diameter; ♀ inflorescences on peduncles 1–3.5 cm. long, sometimes branched dichotomously under the first cymose head, with 1–2 lateral inflorescences; ♀ heads 0.5–0.9 cm. in diameter, with inconspicuous brown bracts at the base. Male flowers on pedicels ± 1 mm. long; perianth 1–1.3 mm. in diameter, 3-merous, tepals with subapical swelling. Female flowers subsessile; perianth 0.8–1.5 mm. long, with 1 narrowly lanceolate perianth-segment and 2 much shorter lateral ones, the longest one with subapical swelling. Achene ovoid, compressed, ± 1.5 mm. long, brown. Fig. 9/J, p. 32.

TANZANIA. Uluguru Mts., Lukwangule Plateau, Mar. 1953, *Drummond & Hemsley* 1641! & Mkambaku [Mkumbaku] Mt., June 1978, *Thulin & Mhoro* 3187! & SE. slope of Tumbako, Feb. 1973, *Pócs & Lungwecha* 6876/E!
DISTR. T 6; not known elsewhere
HAB. In upland rain-forest, often on moist rocks or along streams; 1300–2050 m.

NOTE. The range of variation in this species is quite noticeable in spite of the limited area of distribution; most of the specimens seen represent well-developed plants up to 80 cm. high, while a few are small and prostrate, with almost subcircular leaves, such as *Thulin & Mhoro* 3187, which, however, clearly belongs here on account of stipule and floral characters.

9. P. usambarensis *Engl.*, P.O.A. C: 163 (1895); Rendle in F.T.A. 6(2): 275 (1917); Hauman in F.C.B. 1: 202 (1948). Type: Tanzania, W. Usambara Mts., Mlalo, *Holst* 78 (B, holo.†); E. Usambara Mts., between Amani and Monga, *Verdcourt* 207 (K, neo.!)

Perennial herb with prostrate and ascending stems, up to 0.4(–0.5) m. high. Stems juicy, glabrous. Leaves of a pair subequal or one with longer petiole than the other; stipules short, persistent but insignificant, broadly triangular to almost rim-like, up to 0.8 mm. long, translucent, whitish or greenish; petiole 0.5–4 cm. long, glabrous or with a few hairs at connection with lamina; lamina very variable in size and shape, from almost circular or broadly ovate to lanceolate or elongate, 1–11 cm. long, 1–4.5 cm. wide, base truncate to broadly cuneate, margin serrate, with 3–15 teeth on each side, apex rounded, acute or acuminate; lateral nerves 3–11 pairs, basal pair reaching 2nd–4th tooth from the apex; glabrous on both sides, or rarely with scattered stiff hairs above and spreading hairs on the nerves below, upper surface with regularly scattered linear cystoliths, lower surface with scattered hydatodes. Inflorescences peduncular heads, usually 2 together in the upper leaf-axils, usually the 2 of a pair of the same sex; ♂ inflorescences globose cymose heads, 0.8–1.3 cm. in diameter, on peduncles 1.5–5.5 cm. long; ♀ inflorescences globose cymose heads, 0.4–0.6 cm. in diameter, on peduncles 0.8–1.4 cm. long, sometimes the inflorescence-axis branched 2–3 times below the primary head, so the inflorescence consists of up to 5 heads (rarely only 2, one above the other, as on an unbranched inflorescence-axis). Male flowers on pedicels 0.8–1.4 mm. long; perianth ± 1 mm. in diameter; perianth 3-merous, tepals with subapical appendage. Female flowers subsessile; perianth ± 1 mm. long, one perianth-segment ± 1 mm. long, with a subapical appendage up to 0.8 mm. long, the 2 lateral perianth-segments very short. Achene ovoid, compressed, ± 0.8 mm. long, brown.

KEY TO INFRASPECIFIC VARIANTS

1. Lamina circular or broadly ovate, broader or almost as
       broad as long   . . . .   c. var. **veronicifolia**
   Lamina longer than broad, ovate, elliptic or lanceolate   2
2. Lamina ovate, clearly widest near the base   . . . . .   a. var. **usambarensis**
   Lamina elliptic or lanceolate, widest near the middle   b. var. **engleri**

a. var. **usambarensis**

Habit as above. Lamina ovate, always longer than broad, apex acute, margin crenate or serrate, teeth blunt or acute, rarely rounded; basal pair of lateral nerves reaching 2nd or 3rd tooth from apex; upper surface glabrous or with a few scattered hairs, lower surface rarely with a few hairs on the nerves. Fig. 9/D, p. 32.

KENYA. Teita Hills, Sept. 1953, *Drummond & Hemsley* 4316! & below peak of Vuria, May 1975, *Friis & Hansen* 2675! & Mbololo hill, Sept. 1970, *Faden* 70/543!
TANZANIA. Lushoto District: Mazumbai, July 1966, *Semsei* 4065! & E. Usambara Mts., Amani, July 1903, *Warnecke* 458!; Iringa District: Mufindi, Apr. 1970, *Paget-Wilkes* 822!
DISTR. K 7; T 3, 5–7; Malawi

HAB.   Rain-forest at intermediate altitudes; 900–2200 m.

NOTE.   Material referred to this taxon all has an ovate lamina, which, however, may vary considerably in size. Small forms, with leaves 1–2.5 cm. long, are represented on the Uluguru Mts. (*Bidgood et al.* 211) and Mt. Rungwe (*Bidgood et al.* 92); forms with extremely small lamina have been collected from the Uluguru Mts. (*E. M. Bruce* 942). The only specimen of *P. usambarensis* collected from outside the Flora area (*Dowsett-Lemaire* 716, Viphya Plateau in N. Malawi) belongs to this variety. For taxonomic delimitation, see discussion under var. *engleri*.

b. var. **engleri** (*Rendle*) *Friis* in K.B. 43: 648 (1988). Type: Tanzania, Kilimanjaro, 1900–2900 m, *Volkens* 1494 (B, holo.†); near Moshi, 2600 m., *Drummond & Hemsley* 1290 (K, neo.!, BR, isoneo.!)

Habit as above. Leaves lanceolate to elliptic, always longer than broad, margin serrate, teeth acute, apex acuminate; basal pair of lateral nerves reaching 3rd or 4th tooth from apex, glabrous on both sides. Fig. 9/E, p. 32.

KENYA.   Naivasha District: Kinangop, Apr. 1938, *Chandler* 2359! & W. slope of Aberdare Mts., Apr. 1922, *T.C.E. & R.E. Fries* 2785!; Mt. Kenya, E. side, Jan. 1922, *T.C.E. & R.E. Fries* 1204!
TANZANIA.   Kilimanjaro, Bismarck Hill, Feb. 1934, *Greenway* 3856!; Lushoto District: Shagayu, July 1957, *Carmichael* 636!; Morogoro District: Uluguru Mts., Lukwangule Plateau, Mar. 1953, *Drummond & Hemsley* 1643!
DISTR.   **K** 3/4, 4; **T** 2, 3, 6; not known elsewhere
HAB.   Upland rain-forest; 1500–2800 m.

SYN.   *P. longipes* Engl., P.O.A. C: 163 (1895), *non* Liebm. (1851). *nom. illegit.* Type as for *P. usambarensis* var. *engleri* (see taxonomic discussion below).
          P. engleri Rendle in F.T.A. 6(2): 275 (1917); Hauman in F.C.B. 1: 201 (1948)
          P. engleri Rendle forma *contracta* Hauman in F.C.B. 1: 201 (1948), *nom. inval.*

NOTE.   Some specimens show transition to var. *usambarensis*, e.g. *Harris et al.* 1173 from the Uluguru Mts., and *Greenway* 6550 from the S. Pare Mts. Otherwise, this taxon seems to be sympatric with var. *usambarensis*, but replaces it at altitudes higher than about 2000 m., except for the Southern Highlands of Tanzania, from where the var. *engleri* seems to be absent. It is difficult to interpret this in a satisfactory formal taxonomy, but because of the morphological overlap and the almost sympatric distribution, the rank of variety has been chosen.

c. var. **veronicifolia** (*Engl.*) *Friis* in K.B. 43: 648 (1988). Type: Tanzania, W. Usambara Mts., Kwa Mshusa, *Holst* 9085 (B, holo.†, K, P, Z (all marked "*Holst* 9885") presumably iso.!)

Usually more slender than the previous varieties. Leaves circular to broadly ovate, as wide as long, or wider, margin serrate to crenate, with blunt to acute teeth, sometimes somewhat rounded, apex broadly acute, blunt or rounded; basal pair of lateral nerves reaching 2nd or 3rd tooth from apex; upper surface glabrous or with a few scattered hairs (rarely very hairy), lower surface sometimes with hairs on nerves. Fig. 9/H, p. 32.

KENYA.   S. slope of Mt. Kenya, Oct. 1979, *V. C. Gilbert* 5785!; Fort Hall District: Thika gorge at the Blue Post Hotel, Feb. 1953, *Drummond & Hemsley* 1232!; Kiambu District: Kikuyu Escarpment Forest, Aug. 1974, *Faden et al.* 74/1334!
TANZANIA.   Lushoto District: Mkuzi to Kifungilo, Apr. 1953, *Drummond & Hemsley* 2232!; Kilosa District: Ukaguru Mts., June 1978, *Thulin & Mhoro* 2964!; Iringa District: Mt. Image, Mar. 1954, *Carmichael* 376!
DISTR.   **K** 4; **T** 3, 6, 7; not known elsewhere
HAB.   Upland rain-forest, on forest floor and epiphytic on lower parts of trunks; 1300–2200 m.

SYN.   *P. veronicifolia* Engl., P.O.A. C: 164 (1895); Rendle in F.T.A. 6(2): 274 (1917); U.K.W.F.: 323 (1974)

NOTE.   Some specimens show transition to var. *usambarensis*, e.g. *Snowden* 646, from Kiambu District: Limuru, and *Napier* 2180, from N. Nyeri District: Nanyuki. The same considerations about overlap, both morphological and geographical, apply to this taxon as for the distinction between var. *usambarensis* and var. *engleri*.

10. **P. microphylla** (*L.*) *Liebm.* in Kgl. Danske Vidensk. Selsk. Skr. 5(2): 302 (1851); Rendle in F.T.A. 6(2): 269 (1917); F.W.T.A., ed. 2,1: 621 (1958); Letouzey in Fl. Cameroun 8: 178 (1968). Type: Jamaica, *P. Browne* in *Linnaean Herbarium* No. 1220.8 (LINN, lecto.!)

Erect or procumbent annual (or perhaps short-lived perennial) herb, up to 30 cm. high, usually much branched and forming mats. Stems juicy, glabrous, rarely with a few scattered hairs. Leaves of a pair very unequal in size; stipules minute, up to 1 mm. long, caducous; petioles 0.5–6 mm. long, glabrous; lamina elliptic to ovate or, more often, obovate, 1–15 mm. long, 0.5–4 mm. wide, base rounded to cuneate, margin entire or slightly crenulate, apex rounded to bluntly obtuse; lateral nerves obscure; smooth and glabrous on both sides, upper surface with very prominent linear cystoliths in the epidermis. Inflorescences small pedunculate to subsessile cymose clusters, 1–2 together in the leaf-axils, 1–4 mm. long, usually with a few ♂ and 5–10 ♀ flowers (or entirely with ♀ flowers), sessile, or on pedicels up to 0.5 mm. long; perianth 0.5–0.8 mm. long. Male flowers usually longest, 3-merous; tepals subequal, with subapical swelling. Female flowers with 1 long perianth-segment and ?2 smaller, lateral ones. Achene ovoid, 0.5–0.8 mm. long, smooth, brown.

TANZANIA.   Lushoto District: Amani, Jan. 1950, *Verdcourt* 18!; Morogoro, Mar. 1953, *Drummond & Hemsley* 1480!; Bagamoyo, July 1970, *Thulin & Mhoro* 497!; Zanzibar I., 1 July 1959, *Wild* 4838!; Pemba I., Chake Chake, 24 Aug. 1929, *Vaughan* 572!
DISTR.   **T** 3, 6; **Z**; **P**; Guinea Bissau, Senegal, Liberia, Sierra Leone, Nigeria, Cameroon, Zaire, Rwanda, Burundi, South Africa (Natal); probably native in tropical Central America, but now widely naturalized in the subtropical and tropical parts of the world
HAB.   Along streams on stone walls and between paving, or a garden weed; 0–± 900 m.

SYN.   *Parietaria microphylla* L., Sp. Pl.: 1492 (1753)
    P. serpyllifolia Poir., Encycl. 5: 16 (1804). Type: Martinique, unknown collector, no. 339 (P-LAM, IDC microfiche no. 605III5!)
    Pilea muscosa Lindl., Collect. Bot., t. 4 (1821). Type: a cultivated plant, not traced
    P. serpyllifolia (Poir.) Wedd. in DC., Prodr. 16(1): 107 (1869)

7. ELATOSTEMA

J. R. & G. Forster, Char. Gen. Pl.: 105 (1776); G.P. 3: 386 (1880); Engl. in E. & P. Pf. 3(1): 109 (1888); G.F.P. 2: 184 (1967), *nom. conserv.*

Erect annual and perennial monoecious herbs. Stems prostrate and ascending, with the regularly arranged leaves forming flattened aerial shoots. Cystoliths linear. Leaves alternate, distichous, sessile or very shortly petiolate, asymmetrical; stipules completely or partly fused, intrapetiolar (however, the presence of vestiges of 2 pairs of stipules indicates that the genus has basically opposite leaves, but one leaf of each pair is rudimentary or totally reduced), stipules of reduced leaf different in size and sometimes also in form from those of extant leaf. Inflorescences unisexual, axillary, sessile (in Flora area), consisting of very densely clustered flowers and surrounded by almost free bracts in the ♂ inflorescences or by somewhat fused bracts in the ♀ inflorescences. Male flowers 4–5-merous, tepals generally with a horn-like dorsal appendage; rudimentary ovary poorly developed or absent. Female flowers with 3–5 very reduced tepals and 3 scale-like staminodes, sometimes sterile; ovary ovoid, with penicillate stigma. Achene ovate, small, ejected by the reflexed staminodes.

About 200 species, often very polymorphic, in the tropics of the Old World. One species has been shown to be apogamous.

A special terminology is practical for describing the two halves of the asymmetrical leaves: the "proximal half" is the usually narrow half nearest to the stem, the "distal half" is the usually wider half away from the stem.

The interpretation of the very small and compact inflorescence of this genus is not yet entirely clear; see Letouzey in Fl. Cameroun 8: 144–145 (1968) for further discussion. The taxonomy of *Elatostema* in tropical Africa is not yet satisfactory.

1. Leaves with more than (16–)25 teeth on distal side of lamina; plant drying a fresh green; leaves often more than 15 cm. long . . . .   *1. E. welwitschii*
    Leaves with less than 25 teeth on each side of lamina; plant drying greyish or almost blackish-green; longest leaf usually less than 15 cm. long   2
2. Longest stipule [usually that of the reduced leaf] (5–)6–9(–12) mm. long; number of teeth on distal side of lamina (12–)16–22; lamina usually glabrous above; diameter of ♀ inflorescence 10–12 mm.; widest bract of ♂ inflorescence 3–5 mm. . . . .   *2. E. paivaeanum*
    Longest stipule (1–)2–4(–5) mm. long; number of teeth on distal side of lamina 5–10(–13), lamina above usually with stiff, scattered hairs; diameter of ♀ inflorescences 4–8 mm., widest bracts of ♂ inflorescences 1–2.5 mm.   *3. E. monticolum*

1. E. welwitschii *Engl.* in E.J. 33: 144 (1902); Rendle in F.T.A. 6(2): 278 (1917); Hauman in F.C.B. 1: 206 (1948); F.W.T.A., ed. 2,1: 620 (1958); Letouzey in Fl. Cameroun 8: 151 (1968); Troupin, Fl. Rwanda 1: 160 (1978). Types: Cameroon, W. of Buea, *Preuss* 607 (B, syn.†, not identical with *Preuss* 607 at K which is *E. monticolum*) & Johann Albrechtshöhe, *Staudt* 839 (B, syn.†, BM, isosyn.!) & São Tomé, *Quintas* 149 (B, syn.†, BM, COI, isosyn.!) & Angola, Delamboa R., *Welwitsch* 6269 (?B, syn.†, BM, isosyn.!) & Golungo Alto, near Cacarambola, *Welwitsch* 6270 (?B, syn.†, BM, isosyn.!)

Perennial herb with creeping rhizomes, erect stems fleshy, up to 50 cm. high or sometimes more, up to ± 0.7 cm. in diameter at base, usually little branched, pilose to hispid. Leaves restricted to the upper half of the stem; stipules of reduced leaf similar in size and shape to those of the extant leaf, lanceolate, 8–10 mm. long, with a prominent nerve, which especially in the stipules of the extant leaves, can be discerned to consist of 2 fused nerves, glabrous, green, with translucent margin; petiole usually very short, never more than 0.5 cm. long; lamina obliquely ovate to elliptic, persistently fresh green to yellowish-green, 5–15 cm. long, proximal half of lamina 0.5–2.5 cm. wide, distal half of lamina 1–3.5 cm. wide, base acuminate on both sides of midnerve, margin serrate in the entire length, with (16–)25–40 teeth on distal half tooth, each terminated by a small tuft of hairs and/or a hydatodal gland, apex long acuminate; larger lateral nerves on the proximal side 5–10, on the distal side 5–8, basal lateral nerve on distal side reaching through half the length of lamina or less; upper surface with a few scattered hairs, linear cystoliths prominent, especially along the nerves, lower surface with stiff hairs on the nerves. Inflorescences unisexual, often ♂ and ♀ ones on the same shoot, sessile or on peduncles up to 3 mm. long, in the axils of the extant leaves, rarely in the axils of the stipules of the reduced leaf. Male inflorescences densely cymose clusters, 0.5–0.8 cm. in diameter; bracts largely ovate, 1–2.5 mm. long, ciliate, bracteoles smaller, lanceolate; ♀ inflorescences with numerous minute flowers placed densely together on the flattened axis, often 4-lobed so as to appear cross-shaped, 10–15 mm. in diameter, bracts ovate to triangular, ± 1.5 mm. long, pilose to pubescent, bracteoles lanceolate, ± 1 mm. long. Male flowers on pedicels 2–3 mm. long; perianth ± 1.5 mm. in diameter, 5-merous; tepals pilose, with a short subapical appendage, up to 0.8 mm. long. Female flowers shortly pedicellate, pedicel lengthening during and/or after anthesis; perianth very reduced, consisting of 3–4 tiny ciliate lobes at the base of the ovary; ovary ± 0.5 mm. long, shortly stipitate, ovoid; stigma penicillate; staminodal scales ± half as long as achene. Achene on a fruiting pedicel ± 1 mm. long, with 4–6 fine longitudinal lines or ridges, 0.5–0.7 mm. long, pale brown. Fig. 10/A, p. 40.

UGANDA. Toro District: Nyakagera, Kibale Forest, Aug. 1970, *Katende* 528!; Ankole District: Bunyaruguru, July 1939, *Purseglove* 890!; Kigezi District: Ishasha Gorge, Mar. 1946, *Purseglove* 2007!
TANZANIA. Kigoma District: Kakombe, July 1959, *Newbould & Harley* 4288!; Morogoro District: Uluguru Mts., Hululu Falls on Mgeta R., Mar. 1953, *Drummond & Hemsley* 1621!; Iringa District: Kidatu, Jan. 1971, *Mhoro* 265!
DISTR. U 2; T 3, 4, 6, 7; west to S. Nigeria, São Tomé, south to Angola and Malawi
HAB. Lowland rain-forest, often along streams, riverine forest, often among rocks; 800–1850 m.

SYN. [*Elatostema sessile* sensu auct., *non* Forster (1776): Henriq. in Bol. Soc. Brot. 10: 163 (1892)]
　　E. henriquesii Engl. in E.J. 33: 125 (1902). Type: São Tomé, *Quintas* 150 (B, holo.†, BM, iso.!)
　　E. kamerunense Engl. in E.J. 33: 125 (1902), *nom. inval.*
　　E. welwitschii Engl. var. *cameroonense* Rendle in J.B. 55: 201 (1917). Types: Cameroon, Mt. Cameroon, W. of Buea, *Preuss* 607 (BM, syn.!, B, isosyn.†) & Johann Albrechtshöhe, *Staudt* 839 (BM, syn.!, B, isosyn.†) & Bioko [Fernando Po], Clarence Peak, *Mann* 632 (K, syn.!)

2. E. paivaeanum *Wedd.* in DC., Prodr. 16(1): 178 (1869); Rendle in F.T.A. 6(2): 279 (1917); Hauman in F.C.B. 1: 207 (1948); F.W.T.A., ed. 2,1: 620 (1958); Letouzey in Fl. Cameroun 8: 147 (1968). Types: Bioko, Clarence Peak, *Mann* 290 & 631 (K, syn.!)

Perennial herb with creeping rhizomes, each ending in one or a few erect stems up to 50 cm. high, occasionally up to 1 m., usually unbranched, pubescent in the upper part. Leaves restricted to the uppermost part of the erect stems; stipules slightly dimorphic, those of the extant leaf slightly shorter and wider than those of the reduced leaf, glabrescent, brownish, transparent when dry, often with traces of 2 midnerves, stipule of reduced leaf linear-lanceolate, (5–)6–9(–12) mm. long, long-acuminate with filiform apex; petiole absent or rarely up to 0.5 cm. long; lamina obliquely and broadly obovate to elliptic, (3.5–)7–12(–14) cm. long, proximal half 0.8–1.5 cm. wide, distal half 1–2.5 cm. wide, base asymmetrically cuneate or the distal side subcordate, margin serrate on the distal side, with (12–)16–22 teeth, on the proximal side fewer, apex acute to acuminate;

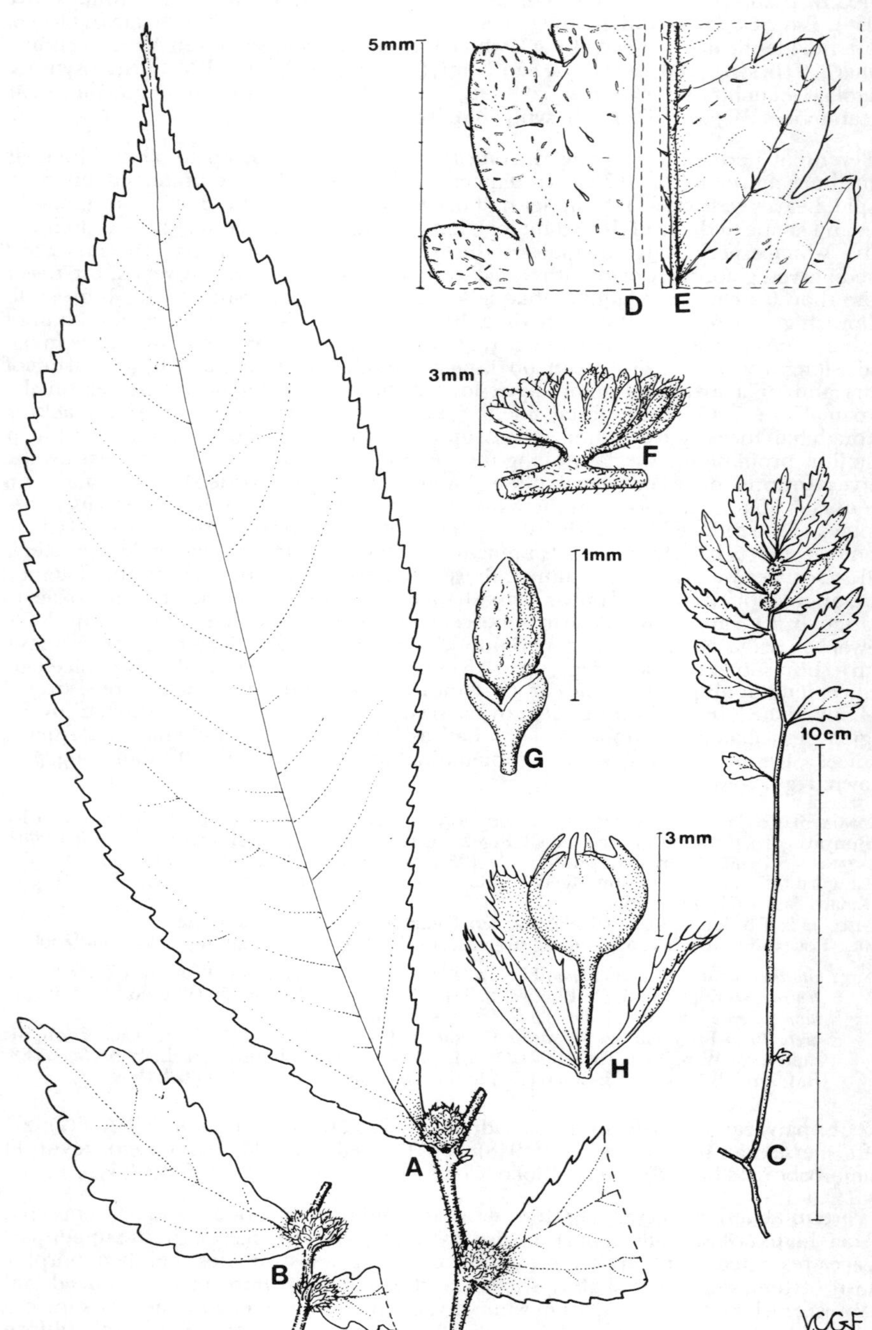

FIG. 10. *ELATOSTEMA WELWITSCHII* — **A**, detail of flowering stem with leaf. *E. PAIVAEANUM* — **B**, detail of flowering stem with leaf. *E. MONTICOLUM* — **C**, habit; **D**, detail of leaf from above; **E**, detail of leaf from below; **F**, inflorescence with female flowers; **G**, old female flower with achene; **H**, male flower in bud. A, from *Grosvenor & Renz* 1227; B, from *Grosvenor & Renz* 1228; C–H, from *Chase* 8152. Drawn by Victoria C. Friis.

prominent lateral nerves 5–6 on the distal side, 3–4 on the proximal side, basal lateral
nerve on distal side reaching less than half through the length of the lamina; upper
surface of lamina usually glabrous, occasionally with a few scattered stiff hairs, numerous
linear cystoliths clearly visible from above, at least in dried material, lower surface
pubescent to strigose on the nerves. Male and ♀ inflorescences frequently on the same
shoot, axillary, sessile or subsessile; ♂ inflorescences up to 1.5 cm. in diameter, bracts
broadly ovate, 3–5 mm. wide, acute to acuminate, ciliate, bracteoles lanceolate, ciliate, ±1
mm. wide; ♀ inflorescences 1–1.2 cm. in diameter, with numerous flowers, bracts
lanceolate, 1–2 mm. wide, ciliate, bracteoles linear-lanceolate, densely pilose, which gives
the entire ♀ inflorescence a greyish appearance. Male flowers on pedicels ± 2 mm. long;
perianth ± 1.5 mm. in diameter, 4-merous, tepals glabrescent, with a ± 1 mm. long
subapical appendage. Female flowers on short pedicels, elongating during or after
anthesis; perianth reduced to 3–4 small ciliate lobes, ovary ± 0.8 mm. long, with penicillate
stigma; staminodal scales ± as long as achene. Achene ± 1 mm. long on pedicels, ± 1 mm.
long, pale brown, smooth or with faint, interrupted longitudinal sculpturing or
longitudinal lines. Fig. 10/B.

Tanzania.   Lushoto District: E. Usambara Mts., at bottom of Bomole near Amani Lake, Feb. 1950,
   Verdcourt 89!; Morogoro District: Uluguru Mts., Lukwangule Plateau, above Chenzema Mission,
   Mar. 1953, *Drummond & Hemsley* 1522! & Uluguru Mts., below the Hululu Falls on the Mgeta R.,
   Mar. 1953, *Drummond & Hemsley* 1576!
Distr.   **T** 2, 3, 6, 7; widespread in the wetter parts of tropical Africa, west to Guineé south to Malawi
Hab.   Lowland rain-forest or altitudinal transitional forest, in the moist ground cover, often along
   streams; 900–2100 m.

Syn.   *E. preussii* Engl. in E.J. 33: 126 (1902). Type: Cameroon, W. of Buea, *Preuss* 586 (B, holo.†)
   E. paivaeanum Wedd. var. *conrauanum* Engl. in E.J. 33: 126 (1902). Type: not indicated
   ?*E. busseanum* H. Winkler in E.J. 41: 277 (1908). Type: Cameroon, Neu-Tegel, *Winkler* 177 (B,
   holo.†)
   E. gabonensis H. Schroeter in F.R. 47: 217 (1939). Type: Gabon/Congo, Mayombe, *Thollon* 1261
   (P, holo.)

Note.   Specimens of this taxon with immature inflorescences can be very difficult to separate from
   large specimens of *E. monticolum*. Collectors should take care to gather good flowering material of
   these two taxa. The species may be more widespread in the Flora area than indicated here, as it
   seems to be undercollected. The species has not been recorded with certainty from Kenya or
   Uganda. One specimen from Kenya (*Perdue & Kibuwa* 8397, Jan. 1967) from Mt. Kenya at 2100 m.
   looks much like *E. paivaeanum*, but in absence of inflorescences the identity cannot be
   ascertained. The best vegetative characters for separating *E. paivaeanum* and *E. monticolum* have
   been included in the key. Specimens with well-developed inflorescences can be recognized by the
   broad bracts of the ♂ inflorescences and the comparatively large ♀ inflorescences with very hairy
   bracteoles.

3. E. monticolum *Hook. f.* in J.L.S. 7: 216 (1864), as "*Elatostemma monticola*"; Rendle in
F.T.A. 6(2): 281 (1917); Hauman in F.C.B. 1: 206 (1948); F.W.T.A., ed. 2,1: 620 (1958);
Letouzey in Fl. Cameroun 8: 141 (1968); Troupin, Fl. Rwanda: 160, fig. 32/2A–B (1978).
Type: Cameroon, Mt. Cameroon, *Mann* 2014 (K, holo.!)

Annual or apparently perennial herbs, apparently dioecious, rarely monoecious,
sometimes flowering shortly after germination on erect stems, 5–10 cm. tall, fixed to the
substratum by fibrous roots, or on ascending stems from a prostrate rhizome, up to 30 cm.
high, pubescent to densely pilose on the upper part of the stem. Leaves mainly on the
upper half of the stems; stipules lanceolate, the ones of the reduced leaf rather like those
of the extant leaf, but separated less than 180° from them on the stem, (1–)2–4(–5) mm.
long, persistent; petiole usually almost absent, rarely up to 0.3 mm. long, pubescent;
lamina obovate to lanceolate, 1.5–6(–8.5) cm. long, proximal half 0.3–1.2 cm. wide, distal
half 0.5–1.5 cm. wide, base very oblique, on proximal side cuneate, on distal side
auriculate or subcordate, margin serrate, distal margin with 5–10 teeth, proximal side with
fewer, apex rounded, acute or acuminate, sometimes caudate; prominent lateral nerves
on the distal side 3–4, on the proximal side 2–4, basal lateral nerve reaching through
more than half the length of lamina before arching towards the midnerve; above with
scattered stiff hairs, rarely glabrous, numerous linear cystoliths clearly visible, lower
surface with stiff hairs on the nerves. Inflorescences sessile, unisexual, axillary,
sometimes contiguous near the top of the stem; ♂ inflorescence ± 0.5(–0.8) cm. in
diameter; bracts lanceolate, obtuse, 1–2(–2.5) mm. wide, ciliate, bracteoles smaller, ciliate;
♀ inflorescences ± 0.5 cm. in diameter (or ?more), bracts lanceolate to oblong, ± 1 mm.
wide, sparsely ciliate, bracteoles linear-spathulate, ciliate in upper half. Male flowers on

pedicels 1.5–2 mm. long; perianth ± 1.5 mm. in diameter, (3–)4-merous, tepals glabrescent, with subapical appendage 0.5–1.5 mm. long. Female flowers probably sessile or subsessile at first, later with elongated pedicel; 3 perianth-segments, extremely reduced; hyaline staminodal scales 3; ovary with linear stripes or fine ridges or almost smooth. Fig. 10/C–H.

UGANDA. Kigezi District: Ishasha Gorge, Nov. 1946, *Purseglove* 2280!; Mbale District: Bulago, Aug. 1932, *A. S. Thomas* 359!; Mengo District: Mpanga Forest, near Mpigi, Nov. 1952, *Dawkins* 761!
KENYA. S. Nyeri District: Kirinyaga, Thiba Fishing Camp, July 1977, *Gilbert & Rankin* 4816!; Kiambu District: Gatamayu [Katamayu] Forest, near Limuru, Sept. 1971, *Verdcourt* 614!; Teita District: Mbololo Hill, Sept. 1970, *Faden et al.* 70/544!
TANZANIA. Kilosa District: Mandege, Apr. 1966, *S. A. Robertson* 152!; Morogoro District: Uluguru Mts., about Morningside, Mar. 1975, *Hooper & Townsend* 916!; Songea District: Matengo Hills, Lupembe Hills, May 1956, *Milne-Redhead & Taylor* 10526!
DISTR. U 2–4; **K** 1–5, 7; **T** 2, 3, 6–8; Cameroon, Bioko [Fernando Po], E. Zaire, Rwanda, Burundi, Sudan (Imatong Mts.), Ethiopia, Zimbabwe
HAB. Upland rain-forest, in moist vegetation on forest floor, often also in bamboo forest, especially along streams; (1200–)1600–2800 m.

SYN. *Elatostemma orientale* Engl., P.O.A. C: 164 (1895); Rendle in F.T.A. 6(2): 280 (1917); Hauman in F.C.B. 1: 206 (1948); U.K.W.F.: 323 (1974). Type: Tanzania, Kilimanjaro, 2200 m., *Volkens* 1259 (B, holo.†, BM, iso.!)
  E. orientale Engl. var. *longiacuminatum* De Wild., Pl. Bequaert. 5: 383 (1932). Type: E. Zaire, Ruwenzori, Lanuri, *Bequaert* 1919 (BR, holo.!)
  E. longiacuminatum (De Wild.) Hauman in F.C.B. 1: 207 (1948)

NOTE. The circumscription of this taxon has been chosen with some hesitation. Not only is it difficult on incomplete material to distinguish between *E. monticolum* and *E. paivaeanum*, but it is also complicated to decide whether to include the West African material (true "*E. monticolum*"), which tends to have blunter leaves, with the East African material (often referred to as "*E. orientale*" or "*E. longiacuminatum*"). And there is also the problem of whether the apparently annual, early-flowering, dwarfed specimens are conspecific with the apparently perennial material with prostrate rhizomatous bases. The author has chosen to consider the West African taxon as representing only part of the range of the variation in the East African taxon, a view also implied by Lambinon in B.S.B.B. 91: 209 (1959) and Letouzey in Fl. Cameroun 8: 145–146 (1968). On the contrary, it has not been possible, with the material available, to maintain the distinction between *E. orientale* and *E. longiacuminatum*, in spite of the distinctions suggested by Lambinon in a key to the two taxa; the discontinuity indicated in the key is not present in the material available for this Flora. One collection (*G. Taylor* 1996, from Mt. Sabinio at 2800 m.) has very variable leaf-morphology within the material represented in the duplicates, with leaves of both typical "*monticolum*" and of typical "*longiacuminatum*" shape. The same applies even to the type material of *E. orientale* (*Volkens* 1269 from Kilimanjaro). The most difficult problem is the status of the annual, dwarfed specimens; however, it seems only possible to explain the very scattered distribution of these forms (of which only one collection, *Hepper & Field* 5413, from the crater lake at Lusanje in the Rungwe District of Tanzania, is from the Flora area) by the assumption that they are habitat-induced modifications of *E. monticolum*. Less dwarfed flowering specimens, with somewhat prostrate stems, but still with the cotyledons extant, have been seen from Cameroon; such transitional forms appear to support the taxonomic treatment adopted here

8. PROCRIS

A. L. Juss., Gen. Pl.: 403 (1789); G.P. 3: 386 (1880); Engl. in E. & P. Pf. 3(1): 109 (1888); H. Schroeter in F.R. 45: 179–192 & 257–300 (1938); G.F.P. 2: 184 (1967)

Perennial herbs, sometimes shrubs, sometimes epiphytic, stems often rather juicy or succulent, monoecious or dioecious by abortion. Leaves basically opposite, but one leaf of each pair is either reduced or early falling, thus the plants appear to have alternate leaves; lamina very unequal-sided at base, entire, crenate or serrate, rather fleshy; stipules intrapetiolar, fused to the apex. Male inflorescences small pedunculate clusters; ♀ inflorescences capitate, with a fleshy subglobose receptacle on short fleshy peduncles, in the axils of fallen leaves. Male flowers 4–5-merous; rudimentary ovary present, obovoid. Female flowers 4–5-merous, with very short perianth; ovary straight, exceeding the perianth; stigma sessile, penicillate. Achene ovate, partly covered by the fleshy persistent perianth.

About 30 species, distributed in the tropics of the Old World.

P. crenata *C. Robinson* in Philipp. Journ. Sci., Bot., 5: 507 (1911); F.W.T.A., ed. 2, 1: 620 (1958); Letouzey in Fl. Cameroun 8: 155 (1968). Type: Philippines, Luzon, Bontoe District,

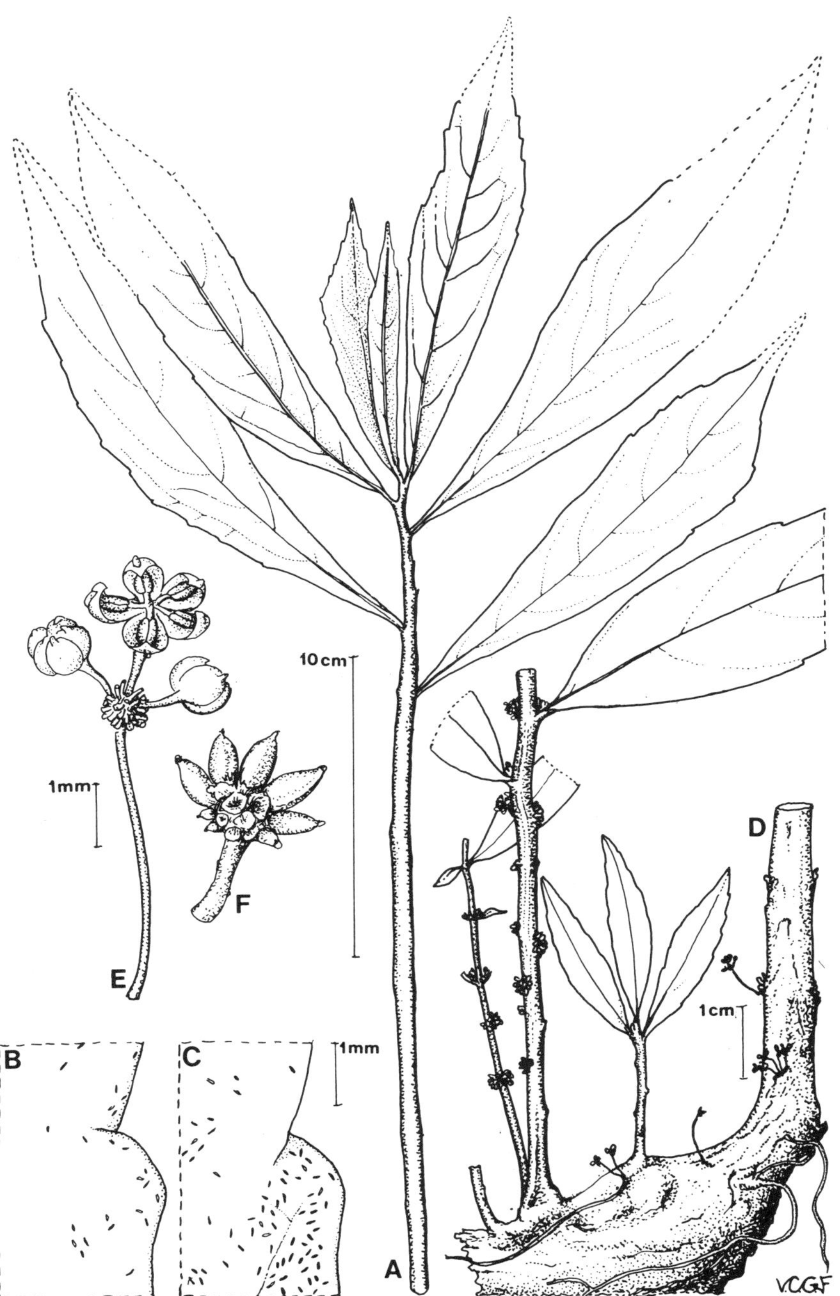

FIG. 11. *PROCRIS CRENATA* — **A**, stem with leaves; **B**, leaf, detail of upper surface; **C**, leaf, detail of lower surface; **D**, basal part of stem; **E**, male inflorescence; **F**, female inflorescence with older flowers. A–C, from *Mueller* 3319; D, E, from Flore du Cameroun; F, from *Banda* 653. Drawn by Victoria C. Friis.

Banco, *Vanoverberg* 635 (?PNH, holo., not seen, B, K, L, NY, iso., according to H. Schroeter)

Perennial herbaceous plants with short, thick and fleshy rhizomatous part, 2 cm. in diameter or more; reported to be monoecious, but no material to confirm this has been seen. Stems fleshy, 1–2 cm. thick when fresh, but shrinking to 4–5 mm. when dry, glabrous, unbranched, arising from the thick rhizomes, reaching a height of 50 cm. or more. Leaves of a pair very unequal, soon falling, the smaller one usually long before the larger one; stipules completely fused, broadly triangular, 1–2 mm. long, obtuse; petiole very short, rarely up to 5 mm. long, passing gradually into the cuneate leaf-base; lamina lanceolate to oblanceolate, widest near or slightly above the middle, slightly asymmetrical, small leaves usually less than 0.8 cm. long and 0.2 m. wide, large leaves 7–15 cm. long, 1–2.5 cm. wide, base long-cuneate, slightly asymmetrical, margin serrate to crenate, rarely subentire, with up to ± 8 poorly marked teeth or crenations, apex long acuminate; lateral nerves in 6–10 pairs, not opposite, almost equally strongly developed; glabrous on both sides, cystoliths elongate, visible on both sides. Male inflorescences (rare in herbarium material and not seen from Flora area) consist of a single few-flowered cluster, up to 0.5 cm. in diameter, on a peduncle 8–10 mm. long; ♀ inflorescences consist of clusters on peduncles 1–5 mm. long, 1–2 mm. thick, each cluster 2–4 mm. in diameter. Male flowers on pedicels 1–2 mm. long; perianth ± 1 mm. in diameter, (4–)5-merous, with corniculate tepals. Female flowers sessile; perianth 4–5-merous, tepals subequal, persistent; ovary erect, with a sessile penicillate stigma. Achene ovoid, hardly compressed, ± 1 mm. long, partly enclosed in the persistent perianth. Fig. 11.

UGANDA.   Kigezi District: Ishasha Gorge, Nov. 1946, *Purseglove* 2287!; Masaka District: Malabigambo Forest, SSW. of Katera, Oct. 1953, *Drummond & Hemsley* 4607!; Mengo District: Mabira Forest, June 1950, *Dawkins* 588!
TANZANIA.   Lushoto District: Vugiri, June 1963, *Archbold* 261!; Morogoro District: Nguru Mts., NW. slope of Mkobwe, Mar. 1953, *Drummond & Hemsley* 1880!; Iringa District: Mufindi, Kigogo R., Mar. 1962, *Polhill & Paulo* 1817!
DISTR.   U 2, 4; T 3, 6, 7; Guineé, Cameroon, Bioko [Fernando Po], Principe, São Tomé, Annabon, Zaire, Angola, Madagascar; tropical Asia from India as far as the Philippines
HAB.   In rain-forest at low or medium altitude, epiphyte on trunks and larger branches in deep shade, less common on the ground or epilithic; 800–1900 m.

SYN.   [*P. wightiana* sensu Rendle in F.T.A. 6(2): 283 (1917); H. Schroeter in F.R. 45: 191 (1938); Hauman in F.C.B. 1: 208 (1948), *non* Wedd. (1856), *nom. illegit.*]

9. BOEHMERIA

Jacq., Enum. Pl. Carib.: 9 & 31 (1760); G.P. 3: 387 (1880); Engl. in E. & P. Pf. 3(1): 111 (1888); G.F.P. 2: 188 (1967)

Small trees or shrubs, sometimes woody-based herbaceous perennials, monoecious or dioecious, usually ± pubescent. Leaves alternate or opposite, equal- or unequal-sided, dentate, sometimes lobed; stipules free, lateral or ± connate. Inflorescences cymose or racemose, with nearly always unisexual clusters of flowers, sessile along the axes of the inflorescence, bracts small, deciduous. Male flowers (3–)4(–5)-merous; rudimentary ovary present. Female flowers with tubular perianth, often contracted at apex, with 2–4 teeth, enclosing the ovary, of which only the filiform, persistent stigma protrudes. Achene enclosed in the persistent membranous perianth.

About 100 species, or probably less, widely distributed in tropical and warm temperate regions. *B. nivea* (L.) Gaud. has been cultivated at Kampala (*Snowden* 1786, Sept. 1930). It is also once been recorded as an escape (*Dummer* 2682, Nov. 1915, at Kigude, Uganda).

B. macrophylla *Hornem.*, Hort. Reg. Bot. Hafn. 2: 890 (1815); Friis & Marais in K.B. 37: 164 (1982); Marais in Fl. Masc., 161, Urtic.: 28 (1985). Type: a plant of unknown origin cultivated in the Botanical Garden of Copenhagen (C, holo.!)

Shrubby or herbaceous perennial plant, monoecious, up to 2–3 m. high, with rather thick unbranched stems, or with long, often overhanging, thinner branches from the underground parts, rarely a small, much branched shrub with slender woody stems; wood of stems juicy or firm. Branches glabrous, pubescent or tomentose, especially in the terminal parts. Leaves opposite, sometimes the 2 of a pair of unequal size; stipules lateral,

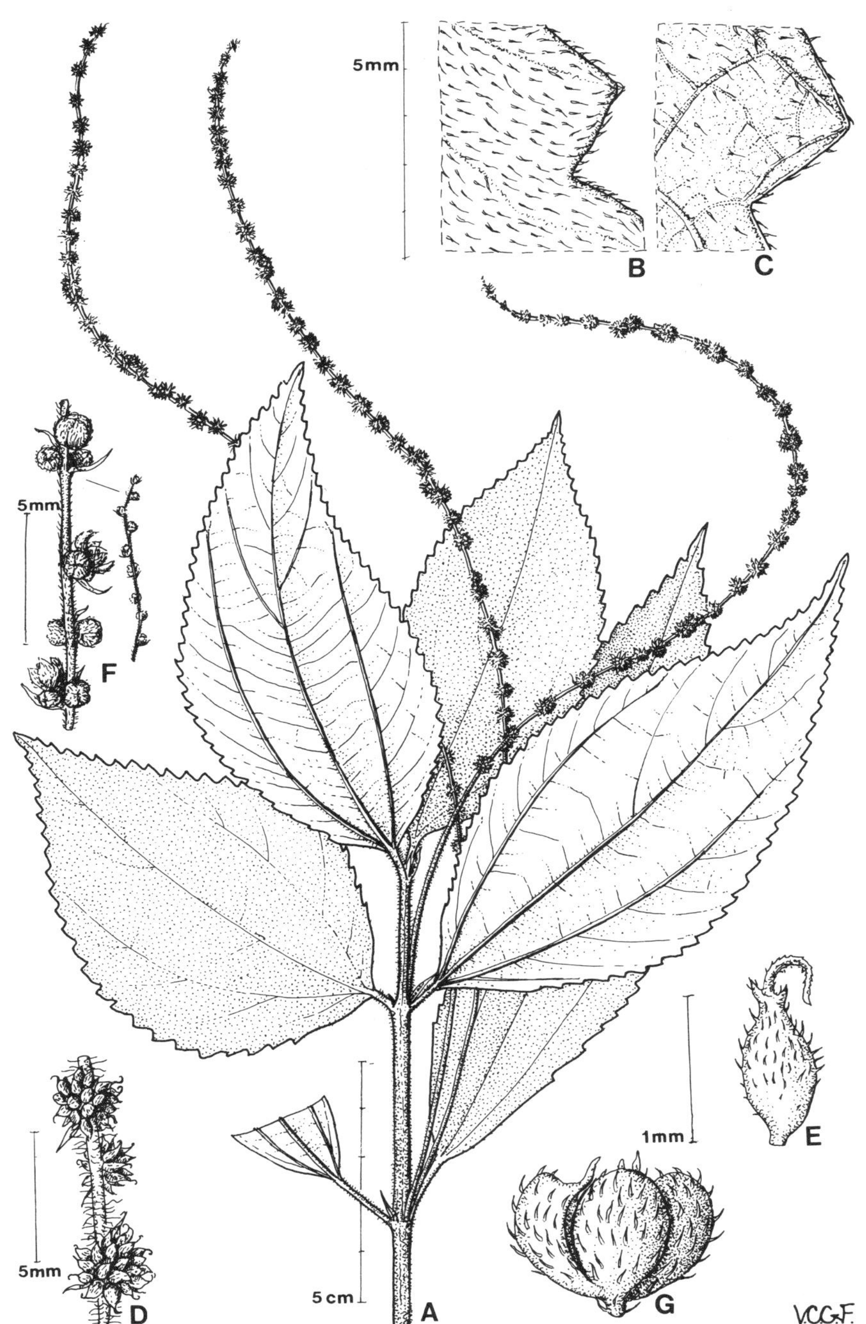

FIG. 12. *BOEHMERIA MACROPHYLLA* — **A**, flowering stem; **B**, detail of upper surface of leaf; **C**, detail of lower surface of leaf; **D**, part of inflorescence; **E**, old female flower with achene enclosed in persisting perianth; **F**, part of male inflorescence; **G**, bud of male flower. All from *Brummitt et al.* 13906. Drawn by Victoria C. Friis.

membranous, brown, up to ± 8 mm. long, lanceolate to narrowly triangular, puberulous on the nerves; petiole 2.5–15 cm. long, glabrous, pubescent or tomentose; lamina ovate to lanceolate, 4–20(–25) cm. long, 2.5–12(–15) cm. wide, base subcordate, truncate, rounded or cuneate, margin serrate, with 15–35 teeth on each side, sometimes without teeth in the basal ⅓, apex acuminate, sometimes almost caudate, tip up to 1.5 cm. long; lateral nerves 3–7 pairs, basal pair prominent, reaching the terminal ⅓ part of lamina before arching towards the midnerve and anastomosing with the upper lateral nerves, rarely ending in the 5th–10th tooth from apex; upper surface glabrescent, scabrid, pubescent or tomentose, with punctiform cystoliths visible, lower surface glabrescent to tomentose. Inflorescences unisexual, rarely with a few flowers of the other sex intermixed, single in the upper leaf-axils, appearing as interrupted spikes, 5–50 cm. long, with the flowers in sessile, dense clusters, spaced 2–6 mm., each young glomerule supported by a caducous bract. Male flowers in glomerules 2–4 mm. in diameter, sessile or subsessile; perianth 4-merous, ± 1 mm. in diameter, tepals with a ± pronounced dorsal gibbosity, apiculate, often with hooked hairs. Female flowers in glomerules 3–5 mm. in diameter, more in fruit, sessile; tubular perianth glabrescent or pubescent, with 2 clearly marked apical teeth, ± 1 mm. long; stigma 1–2.5 mm. long. Achene enclosed in the persistent perianth, ± 2 mm. long, glabrescent to pubescent. Fig. 12.

UGANDA. Bunyoro District: Waki valley, May 1943, *Purseglove* 1581!; Ankole District: Bunyaruguru, Nov. 1938, *Purseglove* 467!; Kigezi District: Ishasha Gorge, Nov. 1946, *Purseglove* 2282!
KENYA. N. Kavirondo District: Kakamega Forest, W. of Forest Station, Apr. 1975, *Friis & Hansen* 2599! & Sept. 1949, *Maas Geesteranus* 6260! & Malaba Forest, Mar. 1969, *Tweedie* 3623!
TANZANIA. Lushoto District: W. Usambara Mts., Jaegertal valley NE. of Lushoto, June 1953, *Drummond & Hemsley* 2969!; Buha District: Kakombe, July 1959, *Newbould & Harley* 4279!; Morogoro District: S. Uluguru Forest Reserve, Lukwangule Plateau, Mar. 1953, *Drummond & Hemsley* 1635!
DISTR. U 2, 4; K 5; T ?1, 3, 4, 6, 7; widespread in tropical Africa in humid lowland areas, west to Guineé, north to SW. Ethiopia, south to Angola, Zimbabwe and Mozambique, also on Madagascar and in tropical Asia as far east as SW. China
HAB. Lowland and upland rain-forest, often in dense tangle along streams, in the undergrowth of riverine forest or among rocks in moist woodland; 800–2150 m.

SYN. *Boehmeria platyphylla* D. Don, Prodr. Fl. Nepal.: 60 (1825); Rendle in F.T.A. 6(2): 285 (1917); Hauman in F.C.B. 1: 210 (1948); F.W.T.A., ed. 2,1: 662 (1958); Letouzey in Fl. Cameroun 8: 196 (1968); U.K.W.F.: 325 (1974); Troupin, Fl. Rwanda 1: 160, fig. 32/4A–C (1978). Type: Nepal, *Hamilton* (BM, holo.!)
　B. platyphylla D. Don var. *angolensis* Rendle in J.B. 55: 201 (1917) & in F.T.A. 6(2): 286 (1917); Hauman in F.C.B. 1: 211 (1948). Types: Angola, Granja de S. Luiz, Cazengo, *Gossweiler* 4656 & 4851 (both BM, syn.!)
　B. platyphylla D. Don var. *ugandensis* Rendle in J.B. 55: 202 (1917) & in F.T.A. 6(2): 286 (1917). Type: Uganda, Ankole, *Dawe* 423 (K, syn.!) & *Scott Elliot* 7531 (BM, syn.!, K, isosyn.!) & Mengo District, Entebbe, *Bagshawe* 799 (?BM, syn.) & Mawokota, *E. Brown* 204 (?BM, syn.) & Kirerema, *Dummer* 89 (BM, syn.!, K, isosyn.!) & Sudan, Monbuttu, Kussumbo R., *Schweinfurth* 3204 (K, syn.!)

NOTE. The record from T1 is based on specimens (*Conrads* 89 & *Stuhlmann* 1148) reported by Peter in F.D.O.-A. 2: 131–132 (1932); specimens from that area have not been seen. The very considerable variation in *B. macrophylla* seems to be somewhat correlated with ecological conditions. Yet it has not been thought useful to create formal taxonomic names for the often very vaguely defined forms. Instead a survey of the major types is given here.
　Very tomentose forms occur in S. & SW. Tanzania and in Uganda under fairly open conditions (e.g. *Cribb et al.* 11294 and *Hepper et al.* 5375, both from Rungwe District, T7, *Procter* 600 from Kibondo, T4, and *Dummer* 3191 from Kivuvu, U4). Forms with smaller leaves, which are very stiff and often somewhat bullose when dry, seem to be very widespread, perhaps in more dry habitats than the following form (e.g. *Schlieben* 1634 from Mahenge, T6, *Cree* 156 from Uganda, and *Dummer* 89 from Kirerema, U4). Forms with large, thin leaves seem to occur in humid forest at low altitude throughout the Flora area (e.g. *Warnecke* in *Herb. Amani* 435 from E. Usambara Mts., T3, *Shabani* 819 from Lushoto, T3, and *Purseglove* 2061 from Kigezi, U2). Forms with rather small, thin leaves, and often much more branched stems than usual occur at higher altitude in forest (e.g. *E.M. Bruce* 620 from Uluguru Mts., T6, *Schlieben* 12267 from Nguru Mts., T6, and *Thulin & Mhoro* 2743 from Ukaguru Mts., T6).

10. POUZOLZIA

Gaudich. in Freyc., Voy. Monde, Bot.: 503 (1830); G.P. 3: 387 (1880); Engl. in E. & P. Pf. 3(1): 112 (1888); G.F.P. 2: 188 (1967); Friis & Jellis in K.B. 39: 587–601 (1984)

Erect annual or, more often, perennial herbs or shrubs, monoecious. Leaves alternate, opposite or verticillate, petiolate or sessile, entire or dentate, cystoliths dot-like; stipules free, lateral. Inflorescence of compact axillary bisexual glomerules. Male flowers pedicellate, 4–5-merous; tepals rarely with transversal crests or wings; rudimentary ovary present or absent. Female flowers sessile, with an indefinite number of completely fused tepals enclosing the ovary, sometimes apparently fused with it in the lower part, often with ± clearly marked longitudinal ridges or wings; staminodes absent; stigma sessile, filiform with lateral papillae, deciduous. Achene ovoid, erect, ± compressed, completely enclosed in the persistent membranous perianth.

About 50 species, mainly in the tropics of the Old World.

1. Leaves all entire 2
   At least some leaves serrate or with dentate margins 5
2. Leaves subsessile, opposite (at least the ones below the inflorescence), sometimes in whorls of 3–4; ♂ flowers all 5-merous, with a transverse dorsal crest on the tepals . . . . *6. P. peteri*
   Leaves petiolate, alternate (except for the lowest one or two pairs), ♂ flowers without transverse dorsal crest 3
3. Shrub with woody stems; stipules brown, elongate, glabrescent or puberulous, not markedly ciliate . . . *3. P. mixta*
   Annual or short-lived perennial, woody only at the base; stipules translucent, rarely brown, lanceolate to caudate, ciliate, with cilia mostly more than 0.5 mm. long 4
4. Inflorescence few-flowered, often consisting of 1–2 ♀ and a number of ♂ flowers; floral bracts conspicuous when present, broadly cordate, with rounded apex; bracts and ♂ perianths with hooked hairs; ♀ perianths pubescent to strigose; stigma usually curled up in a globular structure . . . . *1. P. guineensis*
   Inflorescence many-flowered, usually with 4 or more ♀ and 3 or more ♂ flowers; all floral bracts inconspicuously linear-lanceolate, with acuminate apex; bracts and ♂ perianths puberulous, without hooked hairs; ♀ perianths almost glabrous; stigma filiform, straight, never curled up in a globular structure . . . . *2. P. denudata*
5. Slender annual herb, ultimate branches usually less than 1 mm. thick; the first leaves serrate, but later ones ± entire; achene persistently enclosed in perianth . . . *4. P. fadenii*
   Robust perennial herb, ultimate branches 1 mm. thick or more; leaves always serrate; achene brown or white, easily released from the thin ephemeral perianth . . . *5. P. parasitica*

1. P. guineensis *Benth.* in Hook., Niger Fl.: 518 (1849); Rendle in F.T.A. 6(2): 287 (1917); Hauman in F.C.B. 1: 214 (1948); F.W.T.A., ed. 2, 1: 622 (1968); Letouzey in Fl. Cameroun 8: 185 (1968); Friis & Jellis in K.B. 39: 588 (1984). Types: São Tomé, *G. Don* (BM, syn.!) & Fernando Po, *Vogel* (K, syn.!)

Annual or short-lived perennial up to 1(–2) m. tall, lignified at the base, branched from the base or throughout, pubescent in the upper part, mixed with sparse to abundant long erect hairs. Leaves alternate, usually diminishing in size towards the top of the shoots; stipules membranous, translucent, usually greenish, ovate to lanceolate, shortly caudate, ciliate, 2–4 mm. long, ± 1 mm. wide; petioles 0.1–3(–5) cm. long, pubescent, occasionally with scattered long hairs intermixed; lamina lanceolate to ovate, 1.5–9.5 cm. long, 0.5–3.5 cm. wide, base cuneate, margin entire, apex acuminate; lateral nerves 4–5 pairs, basal pair reaching ¾ towards apex; upper surface with scattered stiff, appressed hairs and punctiform cystoliths, lower surface with a ± prominent arachnoid indumentum (sometimes completely absent), and long stiff hairs on the nerves. Inflorescences axillary, restricted to upper part of the stems, bisexual glomerules, with 1–2(–3) ♀ flowers and a number of ♂ flowers, usually mixed with broadly cordate foliaceous bracts with rounded apex, ± 1 mm. long, densely furnished with hooked hairs (sometimes also with membranous lanceolate, pubescent bracts, perhaps stipules of the cordate bracts). Male flowers on pedicels ± 0.5 mm. long; perianth globular, ± 1 mm. in diameter, pubescent and densely furnished with hooked hairs, 4(–5)-merous. Female flowers sessile; perianth ± 1 mm. long, tubular, constricted at the apex, with 8–10 ± clearly marked longitudinal lines, pubescent to pilose; ovary enclosed, ± 1 mm. long; stigma protruding, bent over or curled up in a glomerule ± 0.5 mm. in diameter. Achene ± 2 mm. long, compressed, white to brown, enclosed in the persistent perianth which is unwinged or rarely provided with 2 very narrow longitudinal wings.

TANZANIA. Buha District: Gombe Stream Reserve, Kakombe valley, Mar. 1964, *Pirozynski* 483!; Kigoma District: Bulimba, May 1975, *Kahurananga et al.* 2651!; Ulanga District: Issongo, Mar. 1932, *Schlieben* 1886!
DISTR. T 4, 6; widely distributed in tropical Africa, from Senegal to S. Sudan and SW. Ethiopia, south to Angola
HAB. Moist wooded grassland, in shade under trees, sometimes among rocks, often associated with the forest-wooded grassland ecotone; 750–1000 m.

SYN. *Parietaria abyssinica* A. Rich., Tent. Fl. Abyss. 2: 258 (1851). Type: Ethiopia, Tigré, Djeladjeranne, in Mai-Mezano valley, *Schimper* 1433 (BM, iso.!)
Margarocarpus schimperianus Wedd. in Ann. Sci. Nat., sér. 4, 1: 205 (1854), *nom. illegit.*, based on *P. abyssinica*
Pouzolzia abyssinica (A.Rich.) Blume in Mus. Bot. Lugd.-Bat. 2: 236 (1856)
P. dewevrei De Wild. & Th. Dur. in B.S.B.B. 38: 53 (Feb. 1900); Rendle in F.T.A. 6(2): 289 (1917). Type: Zaire, Moe, near Bolobo, *Dewèvre* (BR, holo.!)
P. golungensis Hiern, Cat. Afr. Pl. Welw. 1(4): 993 (Aug. 1900). Type: Angola, Golungo Alto, *Welwitsch* 6277 (BM, holo.!, K, iso.!)
P. guineensis Benth. var. *abyssinica* (A. Rich.) Rendle in F.T.A. 6(2): 288 (1917); Hauman in F.C.B. 1: 214 (1948)
NOTE. As this species has been collected from S. Sudan (Imatong Mts.) and SW. Ethiopia, it is very likely also to be found in W. Uganda.

2. P. denudata *De Wild. & Th. Dur.* in B.S.B.B. 38: 54 (1900); Hauman in F.C.B. 1: 213 (1948); Letouzey in Fl. Cameroun 8: 190 (1968); Friis & Jellis in K.B. 39: 589 (1984). Type: Zaire, Kisangani [Stanleyville], Dec. 1896, *Dewèvre* (BR, holo.!)

Perennial herb or subshrub to 2(–3) m. high, woody at the base, little branched, bark greyish brown, young branchlets pubescent. Leaves not or rarely reduced in size towards the tip of branches; stipules ovate, long-acuminate, brown, 5–8 mm. long, ciliate; petioles 1–5.5 cm. long, pubescent, occasionally with long stiff hairs; lamina ovate to lanceolate, 4–11 cm. long, 2–4 cm. wide, base unequal, cuneate, margin entire, apex long-acuminate to caudate; lateral nerves 2–3 pairs, basal pair reaching $\frac{1}{2}$–$\frac{2}{3}$ towards the tip of lamina; upper surface with scattered stiff hairs and punctiform cystoliths, lower surface with scattered stiff hairs on the nerves, sometimes with arachnoid indumentum between the nerves. Inflorescences bisexual, many-flowered, usually with more than 5 ♀ flowers and a similar number of ♂ ones. Male flowers on pedicels ± 0.5 mm. long; perianth globular, 4-merous, pubescent, without hooked hairs. Female flowers sessile; perianth 1–1.5 mm. long, with 3–4 poorly marked teeth at the opening, with 6–10 rather indistinct nerves, glabrous, rarely puberulous towards apex; ovary enclosed; stigma protruding, straight or slightly curved, ± 2 mm. long. Achene slightly compressed, white, brown or black, shiny, 1.5–2 mm. long, enclosed in the persistent perianth.

UGANDA. Bunyoro District: Budongo Forest, Oct. 1971, *Synnott* 721!
DISTR. U 2, ?3; Ivory coast to Zaire, south to Angola
HAB. Lowland rain-forest, often along paths; 900– ± 1050 m.

SYN. *P. andongensis* Hiern, Cat. Afr. Pl. Welw. 1(4): 992 (Aug. 1900); Rendle in F.T.A. 6(2): 289 (1917). Types: Angola, Pungo Andongo, *Welwitsch* 6271 & 6260 (both BM, syn.!, K, isosyn.!)
P. batesii Rendle in J.B. 55: 202 (1917) & in F.T.A. 6(2): 298 (1917). Types: Cameroon, Batanga, *Bates* 214 (BM, syn.!) & Efulen, *Bates* 221 (K, syn.!) & Ngoko, *Schlechter* 12729 (BM, syn.!) & Zaire, Mombongo, *Thonner* 6153 (K, syn.!)
NOTE. One specimen (*Dummer* 5479) from Mabira Forest in Mengo District, Uganda, may also be this species. It is a robust but imperfect specimen which also resembles *P. mixta*, for which species, however, the forest habitat is unlikely.

3. P. mixta *Solms-Laub.* in Sitz. Ges. Nat. Freunde Berlin 1864: 1 (1864) & in Schweinf., Beitr. Fl. Aethiop.: 188 (1867); Rendle in F.T.A. 6(2): 291 (1917); Friis & Jellis in K.B. 39: 590 (1984). Lectotype, chosen by Friis & Jellis (1984): Sudan/Ethiopia, Fazugly, *Cienkowsky* (B, holo.†, LE, lecto.!)

Shrub up to 3(–5) m. high, with soft, juicy, but lignified stems with wide spongy pith or hollow centre, ultimate branchlets more than 1.5 mm. thick; bark longitudinally striate, greyish- or reddish-brown, inflorescence- and leaf-scars prominent, especially on young stems; dense curved or arachnoid hairs on the young parts, later glabrescent. Leaves deciduous, slightly decreasing in size towards tip of the stems; stipules lanceolate, 3–5 mm. long, brown, puberulous, ciliolate; petiole 0.8–1.5(–2) cm. long, densely pubescent; lamina ovate, 2–8(–10) cm. long, 0.8–3(–7) cm. wide, base obliquely cuneate to rounded, margin entire, apex acuminate; lateral nerves 3–4 pairs, basal pair reaching the upper ¾ of lamina; upper surface roughly velvety or scabrid, with scattered to dense stiff hairs and punctiform cystoliths, lower surface densely to slightly arachnoid to white-felted, felt sparse on veins. Inflorescences dense bisexual axillary clusters with short linear bracts, appearing with young leaves, sessile in leaf-axils of current leaves or scattered along twigs in the axils of fallen leaves. Male flowers usually much more numerous than the ♀ ones, on pedicels ± 0.8 mm. long; perianth ± 1 mm. in diameter, (4–)5-merous, tepals densely pubescent, without hooked hairs. Female flowers few, sessile; perianth 1.5–2 mm. long, constricted at the crenulate apex, pubescent, with 3–8 longitudinal ridges, of which 2 are usually larger than the rest; ovary enclosed, except for the 2–4 mm. long straight or slightly curved stigma. Achene ± compressed, smooth, shiny, white to dark brown, 1.5–2.5 mm. long, persistent in and probably sometimes dispersed with the membranous perianth. Fig. 13/A–F, p. 50.

TANZANIA. Mbulu District: Great North Road, Mwembe, Jan. 1962, *Polhill & Paulo* 1956!; Kilosa District: Berega C.M.S. Mission, Dec. 1935, *B.D. Burtt* 5427!; Rungwe Mission, Feb. 1954, *Semsei* 1609!
DISTR. U 2; T 2, ?4,6–8; Ethiopia, ?Sudan, Zambia, Malawi, Mozambique, Zimbabwe, Botswana, SW. Angola, Namibia, South Africa (Transvaal, Natal), also in the Yemen
HAB. Wooded grassland, especially along edges of riverine forest or on rocky outcrops; 100–1200 m.

SYN. *P. hypoleuca* Wedd. in DC., Prodr. 16(1): 227 (1869); Rendle in F.T.A. 6(2): 291 (1917); N. E. Br. in Fl. Cap. 5(2): 551 (1925). Type: Mozambique, Moramballa, *Kirk* (K, holo.!)
  P. arabica Defl., Voy. Yemen: 206 (1889). Type: Yemen, Usil, *Deflers* 215 (P, holo.!)
  P. huillensis Hiern, Cat. Afr. Pl. Welw. 1(4): 993 (Aug. 1900); Rendle in F.T.A. 6(2): 290 (1917). Type: Angola, Huila, Catumba, *Welwitsch* 6290 (BM, holo.!, K, iso.!)
  P. fruticosa Engl. in E.J. 33: 127 (1902); Rendle in F.T.A. 6(2): 292 (1917). Type: Ethiopia, Harar, Erer valley, *Ellenbeck* (B, holo.†)

NOTE. The species has been cultivated at Amani (*Greenway* 2292!) and at Rungwe (*Davies* 272!). The fibre is reportedly used for fishing nets and string. The apparent disjunction of this species from S. Ethiopia to N. Tanzania indicates that it should be looked for in E. Kenya. A record from Uganda (U 2, *Lock* 69/161) has not been seen and may be *P. denudata*.

4. P. fadenii *Friis & Jellis* in K.B. 39: 593 (1984). Type: Kenya, Lamu District, Kitwa Pembe, *Faden* 74/1092 (K, holo.!, MO, iso.!)

Annual herb with thin branched root-system; smaller flowering specimens unbranched, larger ones branching from the base, up to 0.9 m. high; stems glabrescent to pubescent, sometimes with scattered stiff hairs and hooked hairs. Leaves alternate, decreasing in size towards tip of branches; first leaves serrate, later ones entire or with a few basal teeth; stipules lanceolate, 3–5 mm. long, with subcordate base and long-acuminate apex, long ciliate; petiole 0.7–2.5 cm. long, thinly pilose; lamina very variable in size and shape, lanceolate to ovate, 2–8 cm. long, 1–3.5 cm. wide, base cuneate, margin serrate to entire, apex long-acuminate; lateral nerves 2–3 pairs, basal pair reaching the uppermost teeth (or, in entire leaves, the upper ⅘ of lamina); upper surface with scattered stiff hairs and punctiform cystoliths, lower surface glabrous, except for a few scattered hairs on the nerves. Inflorescences bisexual axillary clusters. Male flowers 5–10, on pedicels ± 1 mm. long; perianth ± 1 mm. in diameter, 4-merous, with hooked hairs. Female flowers sessile; perianth ± 1.5 mm. long, with short stiff hairs on apical part and ± 8 rather indistinct longitudinal ridges; ovary enclosed, except for the ± 1 mm. long straight or slightly curved stigma. Achene ± 1.5 mm. long, apparently permanently enclosed in the persistent perianth which becomes slightly verrucose at maturity.

KENYA. Kwale District: Mrima Hill, Sept. 1957, *Verdcourt* 1858!; Kilifi District: Chasimba, Aug. 1971, *Faden* 71/792!; Lamu District: Kitwa Pembe Hill, July 1974, *Faden* 74/1092!
DISTR. K 7; not known elsewhere
HAB. Lowland rain-forest; 0–220 m.

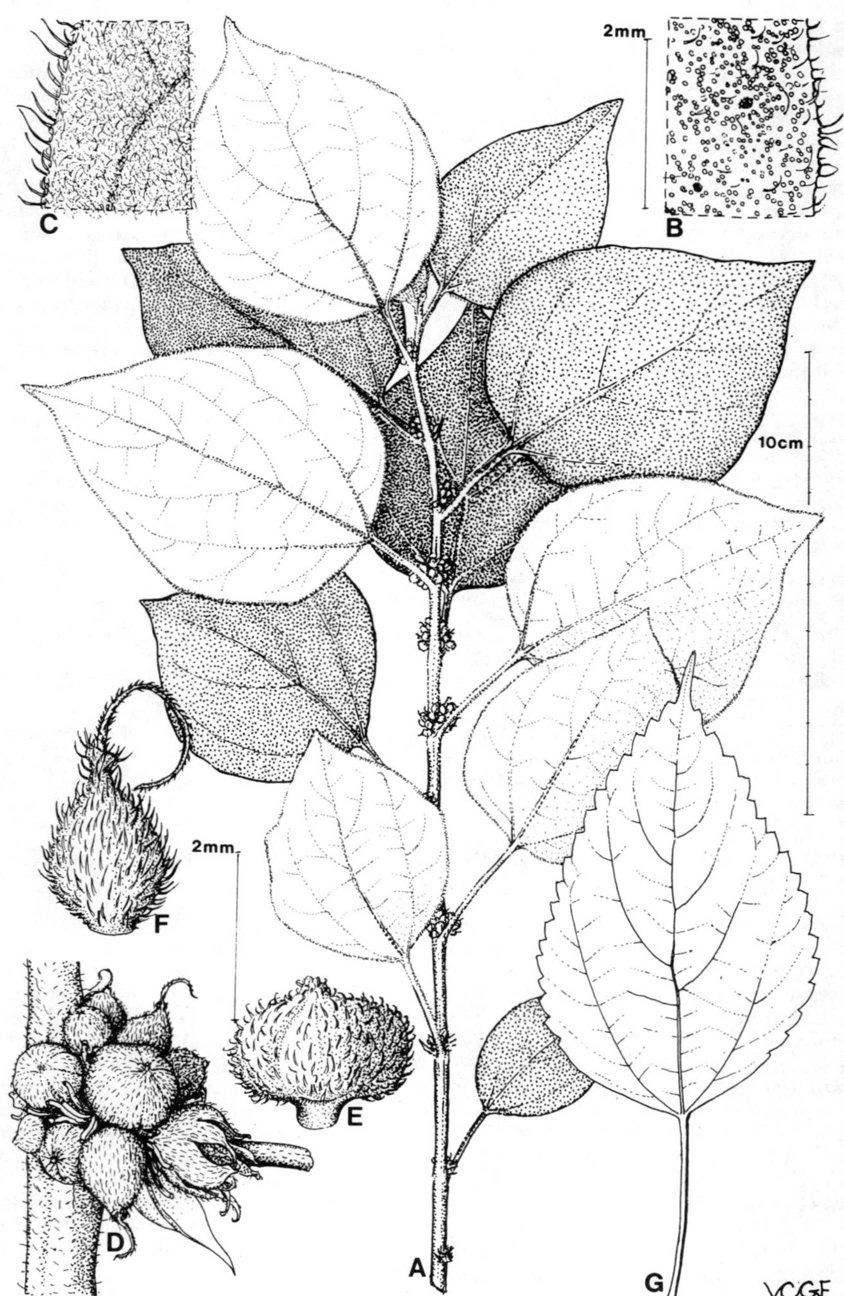

FIG. 13. *POUZOLZIA MIXTA* — **A**, flowering branch; **B**, detail of upper surface of leaf; **C**, detail of lower surface of leaf; **D**, leaf-axil with stipule and inflorescence of male and female flowers; **E**, bud of male flower; **F**, female flower. *P. PARASITICA* — **G**, leaf. A–F, from *Norgrann* 73; G, from *Salubeni & Masinye* 2391. Drawn by Victoria C. Friis.

5. P. parasitica (*Forssk.*) *Schweinf.* in Bull. Herb. Boiss. 4, Appendix 2: 145 (1896); Rendle in F.T.A. 6(2): 293 (1917); Hauman in F.C.B. 1: 215 (1948); F.W.T.A., ed. 2,1: 763 (1958); Letouzey in Fl. Cameroun 8: 194 (1968); U.K.W.F.: 325 (1974). Type: Yemen, Hadie, *Forsskål* (C, holo.!)

Erect or ascending perennial herb up to 1 m. high, sometimes scrambling, often with slender shoots from a woody base, which may be up to 50 mm. in diameter. Stems of ± prostrate forms often rooting at lower nodes; bark brown, pubescent to hirsute or with long dense patent hairs. Leaves alternate; stipules lanceolate, brown, 5–8(–10) mm. long, acuminate, ciliolate, pubescent on midnerve; petiole 2–5(–8) cm. long, pubescent, hirsute or lanate; lamina ovate to lanceolate, 4–7(–10) cm. long, 2.5–4(–6) cm. wide, base rounded to cuneate, margin serrate, with 10–15(–20) teeth on each side, sometimes the lower ⅓ entire, apex acuminate; lateral nerves 3(–4) pairs, basal pair reaching 4th or 5th tooth from apex; upper surface with scattered appressed stiff hairs and punctiform cystoliths, lower surface glabrescent to pubescent, with scattered stiff hairs on nerves. Inflorescences axillary clusters, ± enclosed by the stipules, often appearing as an undeveloped side-shoot, with up to 50 flowers, ♂ flowers more numerous than the ♀ ones. Male flowers on pedicels up to 2 mm. long; perianth ± 1 mm. in diameter, pubescent, sometimes with hooked hairs, (3–)4-merous. Female flowers sessile; perianth 1–2 mm. long, pubescent towards the apex, with 10 very faint longitudinal lines; ovary enclosed, except for the 2–4 mm. long, straight or slightly curved protruding stigma. Achene compressed, white to dark brown, shiny, enclosed in the hairy accrescent perianth, 1.5–2.5 mm. long. Fig. 13/G.

UGANDA. Kigezi District: Kachwekano Farm, Dec. 1951, *Purseglove* 3743!; Mbale District: N. Elgon, Jan. 1953, *Dawkins* 782!; Mengo District: Kipayo, June 1914, *Dummer* 855!
KENYA. Northern Frontier Province: Mt. Kulal, Gatab, Nov. 1978, *Hepper & Jaeger* 6986!; Machakos District: Kibwezi Forest, Apr. 1974, *Faden* 74/425!; Kericho District: SW. Mau Forest, Timbilil, Oct. 1961, *Kerfoot* 2884!
TANZANIA. Mbulu District: Pienaars Heights, Jan. 1962, *Polhill & Paulo* 1093!; Ufipa District: Chala Mt., Dec. 1956, *Richards* 7251!; Songea District: Matengo Hills, Lupembe Hill, May 1956, *Milne-Redhead & Taylor* 10523!
DISTR. U 1–4; K 1–5; T 2–6, 8; Sierra Leone, Cameroon, E. Zaire, Rwanda, Burundi, ?Sudan, Ethiopia, Malawi, Zambia, Zimbabwe, Angola, South Africa (Transvaal, Natal, E. Cape); also Yemen
HAB. Montane rain-forest, especially along paths and streams, and at the edges, riverine forest, in grassland along rivers, between rocks in moist woodland and wooded grassland, especially in S. and C. Tanzania also in *Brachystegia* woodland; 900–2400 m.

SYN. *Urtica parasitica* Forssk., Fl. Aegypt.-Arab.: CXXI, 160 (1775)
U. muralis Vahl, Symb. Bot. 1: 77 (1790), *nom. illegit.*, based on *U. parasitica* Forssk.
Margarocarpus procridioides Wedd. in Ann. Sci. Nat., sér. 4,1: 204 (1854). Types: South Africa, Natal, Pondoland, between Mtata [Umtata] and Umsimubu [St. John's] Rivers, *Drège*, marked "*Urtica ?procridioides* E.M." (P, syn.!, BM, K, isosyn.!) & Umsimubu [St. John's] River, *Drège*, marked "*Urtica ?procridioides* E.M.b." (P, syn.!)
Boehmeria procridioides (Wedd.) Blume in Mus. Bot. Lugd.-Bat. 2: 204 (1856)
Pouzolzia procridioides (Wedd.) Wedd. in Archiv. Mus. Nat. Hist. Nat., Paris 9 (Monogr. Urtic.): 412 (1856); N.E. Br. in Fl. Cap. 5(2): 551 (1925)

NOTE. The specimens of *P. parasitica* from woodland in C. and S. Tanzania have a rather characteristic appearance, being erect plants with many stems from a large woody rootstock, with the stems often only branched at the base and the leaves pointing towards the top of the stems, comparatively short petioles, and elliptic or obovate lamina. This contrasts with the forest forms which are less clearly erect, more diffusely branched, with longer petioles and more clearly ovate lamina. The *Brachystegia* woodland forms also occur in Zambia. However, with the information available, it seems not feasible to decide whether these two forms are mere habitat modifications or represent some kind of genetic separation. They have, therefore, not been given any formal taxonomic status.

6. P. peteri *Friis* in K. B. 42: 666 (1987). Type: plate III/12/2a–i in F.D.O.-A. 2 (1932) (neotype, selected by Friis, l.c.)

Perennial herb to 1.5 m. high, creeping basal parts up to 30 cm. long, with bundles of roots; stems erect, one or a few from the rhizome, branching only in the lower ¼, branches with leaves only; main stem angular, densely pubescent. Leaves in the lower part of the main erect stems opposite, in whorls of 3(–4) or alternate, generally larger than at the tip of the stem, especially at the inflorescence, where they are generally verticillate. Lower cauline leaves subsessile or very shortly pedicellate, ovate to lanceolate, up to 4 cm.

long and 1 cm. wide or wider, rugulose, base cordate to subcordate, margin entire, setulate to serrulate, apex acute to acuminate; lateral nerves ± 2 pairs, basal pair reaching almost the leaf-apex lamina; leaves of branches opposite, lamina broader and shorter than on those of the main stem, apex bluntly acute. Inflorescence bisexual clusters, axillary. Male flowers shortly pedicellate, 5-merous, with a marked transverse dorsal crest on each tepal, bearing a few scattered hairs; rudimentary ovary present. Female flowers: perianth 4-toothed at apex, with 8–10 clearly marked longitudinal ridges; ovary completely enclosed, and only the ± 0.7 mm. long filiform style protruding. Achene shiny black, ovoid, ± 1 mm. long.

TANZANIA. Kigoma District: Ujiji, between the Muguti [Mkuti] R. and Musosi [Mchaji], Feb. 1926, *Peter* 37231; Buha District: between Bugaga [Bugaya] and Kasulu [Kassulo], Feb. 1926, *Peter* 37372 & *Peter* 37415

DISTR. **T** 4; Burundi

HAB. Humid grassland, along streams; 950–1350 m.

SYN. *P. cordata* Peter in F.D.O.-A. 2: 133 (name), & Appendix: 2 (description) & t. II/12/2a–i (1932); Hauman in F.C.B. 1: 216 (1948), *non* Bennett (1838), *nom. illegit.* Types: Kigoma District, Ujiji, Mkuti R.–Mchaji, *Peter* 37231 (B, syn.†) & Buha District, Bugaya–Kasulu [Kassulo], *Peter* 37372 (B, syn. †)

NOTE. All material of this species in herbaria would appear to have been destroyed, including the two specimens designated types of *P. cordata* by Peter, and the only available type for the new name is, therefore, Peter's illustration. From the description and the designation of types it appears that Peter's name is a later homonym of Bennett's *P. cordata*, and clearly not a misapplied name.

 P. peteri belongs to a group of species widely distributed in tropical Asia and not otherwise known from Africa. However, it seems so well documented by Peter that it has been enclosed here on his authority.

11. PARIETARIA

L., Sp. Pl.: 1052 (1753) & Gen. Pl., ed. 5.: 471 (1754); G. P. 3: 392 (1880); Engl. in E. & P. Pf. 3(1): 115 (1888); Hutch., G. F. P. 2: 193 (1967)

Annual or perennial herbs, polygamous. Leaves alternate, petiolate, entire, cystoliths dot-like; stipules absent. Inflorescences of paired sessile or shortly pedunculate clusters in leaf-axils; bracts sometimes very prominent. Flowers bisexual or unisexual. Male and ⚥ flowers: 3- or 4-merous; tepals almost free; rudimentary ovary present in ♂ flowers. Female flowers: with ± tubular, 3- or 4-lobed perianth, stigma sessile, penicillate, decidous, ovary symmetrical. Achene enclosed in persistent perianth.

About 10 species widely distributed in the tropics, subtropics and warm temperate regions; one polymorphic species in mountains of tropical and southern Africa.

P. debilis *G. Forster*, Fl. Ins. Austr. Prodr.: 73 (1786); Rendle in F.T.A. 6, 2: 298 (1917); Hauman in F.C.B. 1: 209 (1948); A.V.P.: 73, 268 (1957); F.W.T.A., ed. 2, 1: 622 (1958); Letouzey in Fl. Cameroun 8: 209 (1968); U.K.W.F.: 325 (1974); Troupin, Fl. Rwanda 1: 158, fig. 32/1A–D (1978). Type: New Zealand, *Sparman & Forster* (?GOET, *fide* Hedberg in A.V.P. and Letouzey l.c.)

Annual, or perhaps sometimes short-lived perennial herb. Stems prostrate, ascending or erect, occasionally scrambling ("*P. laxiflora*" and "*P. scandens*"), delicate or up to 2 m. high, sometimes rooting at the nodes, glabrous, pubescent or with short stiff hairs, occasionally with sessile or stalked reddish glands when young. Leaves alternate, rarely subopposite; stipules lacking; petiole 0.8 – (in "*P. laxiflora*")± 5 cm. long, pubescent to glabrescent; lamina broadly ovate to circular or broader than long, 0.4–10 cm. long, 0.3–4 cm. wide, base often oblique, truncate to slightly cordate, margin entire, apex obtuse or rounded, rounded, obtuse or acuminate; lateral nerves 2–3 pairs, basal pair reaching $\frac{1}{2}$–$\frac{2}{3}$ towards the apex; both surfaces glabrous or upper surface with short stiff hairs, cystoliths punctiform, and lower surface pilose or pubescent, with longer hairs along veins and margin, where long curved hairs may be intermixed. Inflorescence axillary, ± cymosely branched, up to 1.5 cm. long (in "*P. laxiflora*"), with clusters of 1–few sessile or shortly pedicellate flowers, each cluster with ♂ and ♀ intermixed, and bracts with glandular hairs and long curved hairs. Male flowers subsessile; perianth 4-merous, up to 1.5 mm. in diameter. Female flowers shortly pedicellate; perianth 4-merous, ± 1.5 mm. long, with minute hairs on outer surface; ovary enclosed, with protruding penicillate stigma.

FIG. 14. *PARIETARIA DEBILIS* — **1**, flowering and fruiting branch of lax form, × ⅔; **2**, flowering branch of large-leaved form, × 1; **3**, leaf, detail of upper surface, × 6; **4**, detail of inflorescence, × 4; **5, 6**, female flower and fruit, with perianth partly removed, × 14; **7**, branch of slender form, × 1; **8**, hermaphroditic flowers, × 12; **9**, fruit, × 10. Drawn by Hélène Lamourdedieu. Reproduced from Flore du Cameroun.

Bisexual flowers with perianth bearing short hairs on outer surface, ± 1 mm. long. Achene ± 1.5 mm. long, smooth, shiny, dark brown, long enclosed in the persistent, accrescent perianth. Fig. 14.

UGANDA. Toro District: Ruwenzori, Aug. 1938, *Purseglove* 259! & Kibale Forest, July 1938, *A. S. Thomas* 2302!; Kigezi District: Marambo, Kayonza, Mar. 1947, *Purseglove* 2386!
KENYA. Elgon, May 1948, *Hedberg* 925!; Mt. Kenya, Dec. 1957, *Verdcourt* 2011!; Masai District: summit of Ngong Hills, July 1953, *Bally* 9003!
TANZANIA. Ngorongoro Crater, Olmonge valley, Feb. 1962, *Newbould* 5957!; Arusha District: Ngurdoto Crater, Oct. 1965, *Greenway & Kanuri* 11962!; Kilimanjaro, above Peter's Hut, June 1948, *Hedberg* 1205!
DISTR. U 2; K 3, 3/5, 4, 6; T 2; Cameroon, Bioko [Fernando Po], E. Zaire, Rwanda, Burundi, Sudan (Jebel Marra), Ethiopia, Somalia, Namibia, South Africa (E. Orange Free State, Natal Drakensberg), Lesotho; also on Madagascar, in the mountains of tropical Asia, Australia and New Zealand; South America
HAB. In upland rain-forest, dry montane forest, bamboo forest, and evergreen bushland, in shade under rocks and along small streams in montane grassland and upland moor; 1700–4200 m.

SYN. *P. ruwenzoriensis* Cortesi in Ann. Bot. Roma 6: 535 (1908); Rendle in F.T.A. 6(2): 297 (1917). Type: Uganda, Ruwenzori, Mobuku valley, Bujongolo, 3800 m., *Duke of Abruzzi Exped.* (TO, holo, according to Hedberg in A.V.P., l. c.)
P. scandens Engl. in Z.A.E. 1907–1908: 191 (1911). Type: Zaire, Ruwenzori, Butagu valley, 3400 m., *Mildbraed* 2676 (B, holo.†)
P. laxiflora Engl. in Z.A.E. 1907–1908: 191 (1911); Rendle in F.T.A. 6(2): 296 (1917); Hauman in F.C.B. 1: 209 (1948); F.W.T.A., ed. 2,1: 622 (1958). Type: Cameroon, Mt. Cameroon, Mann's Spring, 2200 m., *Mildbraed* 3434 (B, syn.†) & Mfonga, 1700–1900 m., *Ledermann* 5895 (B, syn.†)

NOTE. The East African material of *Parietaria* is rather variable. Large forms have been distinguished as "*P. laxiflora*", "*P. ruwenzoriensis*", and "*P. scandens*", but the view of Hedberg in A.V.P., according to which the variation in size in the tropical African material is due to environmental modifications (and all taxa described from tropical Africa consequently considered conspecific), has been followed here. Recently Hara (in H. Ohashi, ed., The Flora of Eastern Himalaya, Third Report: 23 (1975)) has suggested that the W. Himalayan plants of this taxon should be referred to a separate species, *P. micrantha* Ledeb. It has been suggested (B.L. Burtt in Notes Roy. Bot. Gard. Edinb. 43: 402 (1986)) that the specimens of *Parietaria* from the mountains of tropical and South Africa are conspecific with those from the W. Himalaya, and that all this material should be referred to *P. micrantha* Ledeb. It seems clear that the South African specimens belong to the same species as the W. Himalayan material, but it is also very similar to at least part of the material from New Zealand and Polynesia, the region from which the type of *P. debilis* originated. In absence of a typification of *P. debilis* and a more complete study of the variation of the tropical and Southern Hemisphere species of the genus I am very reluctant to change the name for the African species, and have, therefore, retained it as *P. debilis*.

12. FORSSKAOLEA

L., Diss. Opobals. Decl.: 17 (1764); G.P. 3: 393 (1880); Engl. in E. & P. Pf. 3(1): 117 (1888); G.F.P. 2: 194 (1967); Friis & Wilmot-Dear in Nordic Journ. Bot. 8: 34 (1988)

Annual or perennial herbs or shrublets, monoecious, often covered with hispid hairs. Leaves alternate, petiolate, variously serrate; cystoliths dot-like, prominent; stipules lateral, free. Inflorescences bisexual, rarely unisexual (♀), sessile in leaf-axils, enclosed in a campanulate involucre of bracts; bracts almost completely covered by long dense woolly indumentum (but hairs at base of bracts long and stiff), lanceolate, ovate or obovate, free or fused at base. Male flowers pedicellate, mostly near edge of inflorescence; perianth irregularly 3-lobed; stamen 1, inflexed, later reflexing; rudimentary ovary absent. Female flowers few at centre of inflorescence, without perianth; ovary erect; stigma sessile and filiform. Achene ovate, covered by woolly indumentum.

A genus of 5–6 species occurring in drier parts of Africa, on the Macaronesian Islands, in Spain and extending to Pakistan and NW. India. All species occur in dry areas, and are taxonomically difficult. Merxmueller & Roessler in Misc. Pap. Landb. Wageningen 19: 263–280 (1980) have shown that in Namibia, where most of the African species occur, there are frequently intermediate specimens, but that these intermediate specimens are localized to comparatively restricted transition zones, a phenomenon which is explained by assuming introgressive hybridization.

F. viridis *Webb* in Hook., Niger Fl.: 179 (1849); Rendle in F.T.A. 6(2): 302 (1917); Friis in Taxon 31: 729 (1982). Lectotype, chosen by Friis (1982): plant grown at the Paris Botanical

FIG. 15. *FORSSKAOLEA VIRIDIS* — **1**, habit, × ²/₃; **2**, detail of leaf to show indumentum of upper and lower surface, × 6; **3**, **4**, side and top view of bisexual involucre, × 4; **5**, male flower, × 4; **6**, female involucre, × 8; **7**, female flower, × 8. All from *Gilbert & Thulin* 196. Drawn by Eleanor Catherine. Reproduced with modification from Flora of Ethiopia.

Garden from seed collected on the Red Sea coast of Ethiopia by *Ehrenberg* (FI-WEBB, lecto.!)

Annual or short-lived perennial herb up to 1.2 m. high, branched from a usually herbaceous base. Stems with sparse broad-based curved hairs and denser shorter stiff hairs, hairs often hooked. Leaves regularly scattered along the stems, usually largest near the tip of the branches; stipules with ciliate margin, broadly ovate, up to 2 mm. long, 1.5 mm. wide; petiole 13–30(–45) mm. long; lamina lanceolate to ovate, 1–7(–9) cm. long, 0.5–2(–5) cm. wide, apex broadly acute, base decurrent, margin crenate with 4–8 teeth on each side; lateral nerves 3–4 pairs, basal pair reaching 3rd or 4th tooth from apex; upper surface with scattered hooked hairs, lower surface usually with dense white wool (wool thinner on veins which often have curved hairs). Inflorescence with (3–)4–7(–8)-lobed involucre, dimension of lobes as in key, otherwise as for genus. Flowers as for genus. Achenes shortly stipitate, ovoid, 2–3 mm. long, 1.25–2 mm. wide, dark brown. Fig. 15.

KENYA. Northern Frontier Province: Lokori, 11 km. S. of Kangetet, May 1970, *Mathew* 6397! & S. Turkana, foot of Loriu Plateau, R. Lochinimwoi, Sept. 1969, *Mwangangi* 1524!; Baringo District, Lake Baringo, Ol Kokwa I., June 1977, *M. G. Gilbert* 4718!
TANZANIA. Masai District: Engaruka road, Feb. 1970, *Richards* 25478! & Ang'ata Salei, Lolgarien Gorge, Aug. 1962, *Newbould & Russell* 6256!
DISTR. **K** 1, 3; **T** 2; Ethiopia, Sudan, Somalia, S. Egypt, S. Angola, Namibia (north of Windhoek); also in the Yemen, Saudi Arabia, and on the Cape Verde Is.
HAB. On rocky soil with sparse vegetation, often on rocky slopes and at dry watercourses in shade; 300–1300(–1700) m.

SYN. *F. eenii* Rendle in J.B. 55: 203 (1917) & in F.T.A. 6(2): 301 (1917). Type: Namibia, Damaraland, *Een* (BM, holo.!)

NOTE. In Namibia transitional forms to *F. hereroensis* Schinz exist, see Merxmueller & Roessler in Misc. Pap. Landb. Wageningen 19: 269–277 (1980); for a discussion of the identity of the Cape Verde population, see Lobin & Roessler in Senck. Biol. 65: 373–390 (1985). The record of the exceptionally high altitude of 1700 m. (5500 ft.) for the locality at the Lolgarien Gorge should be checked.

13. DROGUETIA

Gaudich. in Freyc., Voy. Monde, Bot.: 505 (1830); G.P. 3: 394 (1880); Engl. in E. & P. Pf. 3(1): 117 (1888); Friis & Wilmot-Dear in Nordic Journ. Bot. 8: 36 (1988)

Didymogyne Wedd. in Ann. Sci. Nat., sér. 4,1: 207 (1854)

Droguetia sect. *Didymogyne* (Wedd.) Benth. in G.P. 3: 394 (1880)

Annual or perennial herbs, occasionally subshrubs, with erect or prostrate stems, usually monoecious (or plants with only ♀ flowers by abortion of all bisexual inflorescences). Leaves opposite or alternate, petiolate; cystoliths punctiform; stipules lateral, free. Flowers in sessile inflorescences, single or few–many together, surrounded by an involucre of fused bracts; inflorescences bisexual or unisexual by abortion (and then usually ♀), axillary and/or in terminal spikes (if each involucre contains only 1(–2) ♀ flowers, the number of involucres in each leaf-axil is usually high). Male flowers ± boat-shaped; perianth cylindrical to conical at the base, open along one side and with a ± erect tip, enclosing 1 at first inflexed, later reflexed stamen; rudimentary ovary absent. Female flowers without perianth, consisting of a pistil only; style filiform. Achenes enclosed in the persistent involucre.

7 species, distributed in tropical and South Africa, Yemen, S. India and Java.

Leaves on the main stems always opposite, leaves on the side-
    branches usually opposite, or a few alternate . . . . *1. D. iners*
At least the uppermost leaves on the stem, or some of the leaves
    on the side-branches regularly alternate . . . . . *2. D. debilis*

1. D. iners (*Forssk.*) *Schweinf.* in Bull. Herb. Boiss. 4, Appendix 2: 146 (1896); Rendle in F.T.A. 6(2): 303 (1917); Hauman in F.C.B. 1: 217 (1948); F.W.T.A., ed. 2,1: 622 (1958); Letouzey in Fl. Cameroun 8: 213 (1968); U.K.W.F.: 325 (1974); Troupin, Fl. Rwanda 1: 161,

Fig. 33/1A–D (1978); Friis & Wilmot-Dear in Nordic Journ. Bot. 8: 38 (1988). Type: Yemen, *Forsskål* (C, holo.!)

Perennial herb or subshrub with prostrate and erect stems, up to 1 m. high or probably more. Branches and petioles with a very variable indumentum, ranging from subglabrous or with scattered stiff hairs or appressed hairs to an almost woolly tomentum of straight or curly, erect hairs. Leaves opposite, very rarely alternate on an occasional side branch; stipules brown, translucent, with dark brown midnerve, lanceolate, with elongated tip, up to 3 mm. long; petiole 1–2.5(–4) cm. long, pubescent to pilose; lamina ovate, 1.5–4.5(–6.5) cm. long, 0.8–2.5(–3) cm. wide, base cuneate, margin serrate, with 6–22 teeth on each side, apex acute or acuminate, with long apical tooth; lateral nerves 2–3(–4) pairs, basal pair reaching 2nd–6th tooth from apex; upper surface with scattered hairs, lower surface with scattered stiff hairs on the nerves and sometimes also between the nerves. Inflorescences axillary, either bisexual, globular and up to 6 mm. in diameter, or entirely ♀, smaller, ovoid, with 1–2 flowers; in the upper leaf-axils often arranged in clusters, either all bisexual or some bisexual, many-flowered and some ♀, or all ♀, few-flowered; in the lower axils often only ♀ ones present. Achene persistent in the involucre, glabrous or lanate.

subsp. **iners**

Stems glabrescent to pubescent or with scattered stiff hairs, never woolly. Involucre glabrescent on inside, not with indumentum forming a white fringe around flowers. Fig. 16/K–U, p. 58.

UGANDA. Kigezi District: Rubaya, July 1945, *A. S. Thomas* 4268! & Kachwekano Farm, Dec. 1949, *Purseglove* 3176! & Mafuga Forest, Apr. 1950, *Dawkins* 580!
KENYA. Trans-Nzoia District: Elgon, Suam R., Dec. 1967, *Mwangangi* 406!; Kiambu District: near Kinari [Kinale], Oct. 1959, *Verdcourt* 2485!; Masai District: Nasampolai [Enesambulai] valley, Dec. 1969, *Greenway & Kanuri* 13882!
TANZANIA. Mbulu District: Pienaars Heights, May 1962, *Polhill & Paulo* 2338!; Iringa District: Mufindi, Livalonge Tea Estate, Aug. 1971, *Perdue & Kibuwa* 11270!; Rungwe District: Mwakaleli, near Mwatesi R., May 1975, *Hepper & Field* 5455!
DISTR. U 1–3; **K** 1, 3–6; **T** 2, 3, 6, 7; occurs through the highlands of eastern tropical Africa from N. Ethiopia, S. Sudan (Imatong Mts.) to Angola and South Africa (Transvaal, Natal, Cape Province), west to the central highlands of Bioko [Fernando Po] and the highlands of Cameroon; also recorded from the highlands of Yemen

HAB. Upland rain-forest and evergreen bushland, in undergrowth, sometimes persisting in hedges, etc. of farmland after forest clearing; 1500–3050 m.

SYN. *Urtica iners* Forssk., Fl. Aegypt.-Arab.: 160 (1775)
 Parietaria capensis Thunb., Prodr. Pl. Cap. 1: 31 (1794). Lectotype, chosen by Friis in Taxon 35: 703 (1986): South Africa, Cape Province, probably Grootvadersbosch, *Herb. Thunberg* 22129 (UPS-THUNB, lecto.!)
 Boehmeria capensis (Thunb.) Spreng., Syst. Veg., ed. 16, 3: 844 (1826)
 Urtica pauciflora Steudel in Flora 33: 258 (1850). Type: Ethiopia, Semien, Mt. Silke, *Schimper* 682 (P [specimen marked "Herb. Steudel"], holo.!, BM, G, K, M, P, S, UPS, iso.!)
 Pouzolzia pauciflora (Steud.) A. Rich., Tent. Fl. Abyss. 2: 259 (1851)
 Didymogyne abyssinica Wedd. in Ann. Sci. Nat., sér. 4,1: 207 (1854), *nom. superfl.*, based on *U. pauciflora* Steud.
 Droguetia diffusa Wedd. in Ann. Sci. Nat., sér. 4,1: 211 (1854). Lectotype, chosen by Friis & Wilmot-Dear (1988): Ethiopia, Semien, Mt. Silke, *Schimper* 682 (P, lecto.!, BM, G, K, M, P, S, UPS, isolecto.!)
 Boehmeria pauciflora (Steud.) Blume, Mus. Bot. Lugd.-Bat. 2: 201 (1856)
 Droguetia pauciflora (Steud.) Wedd. in DC., Prodr. 16(1): 235/58 (1869)
 Urtica urens L. var. *iners* (Forssk.) Wedd. in DC., Prodr. 16(1): 40 (1869)
 Droguetia woodii N.E. Br. in Fl. Cap. 5(2): 561 (1925). Type: South Africa, Natal, Inanda, *J. M. Wood* 1243 (K, holo.!)
 D. thunbergii N. E. Br. in Fl. Cap. 5(2): 558 (1925), *nom. illegit.*, based on the same type as *Parietaria capensis* Thunb.

DISTR. (of species as a whole). The species as a whole occurs from West Africa (in the highlands of Cameroon) and N. Ethiopia through the highlands of eastern Africa to South Africa (Cape Province); also in the highlands of Yemen, S. India and Java. Subsp. *burchellii* (N. E. Br.) Friis & Wilmot-Dear, differing in woolly indumentum on stems and petioles, occurs in the Cape Province, only ecologically separated from the typical subsp., while subsp. *urticoides* (Wight) Friis & Wilmot-Dear, differing in the woolly indumentum on inside of involucre, replaces the typical subsp. in S. India and on Java.

2. D. debilis *Rendle* in J.B. 60: 203 (1917) & in F.T.A. 6(2): 304 (1917); Hauman in F.C.B. 1: 217 (1948); U.K.W.F.: 325 (1974); Friis & Wilmot-Dear in Nordic Journ. Bot. 8: 44, fig. 5A–J (1988). Type: Kenya, Nakuru/Masai District, Mau, *Scott Elliot* 6799 (BM, holo.!)

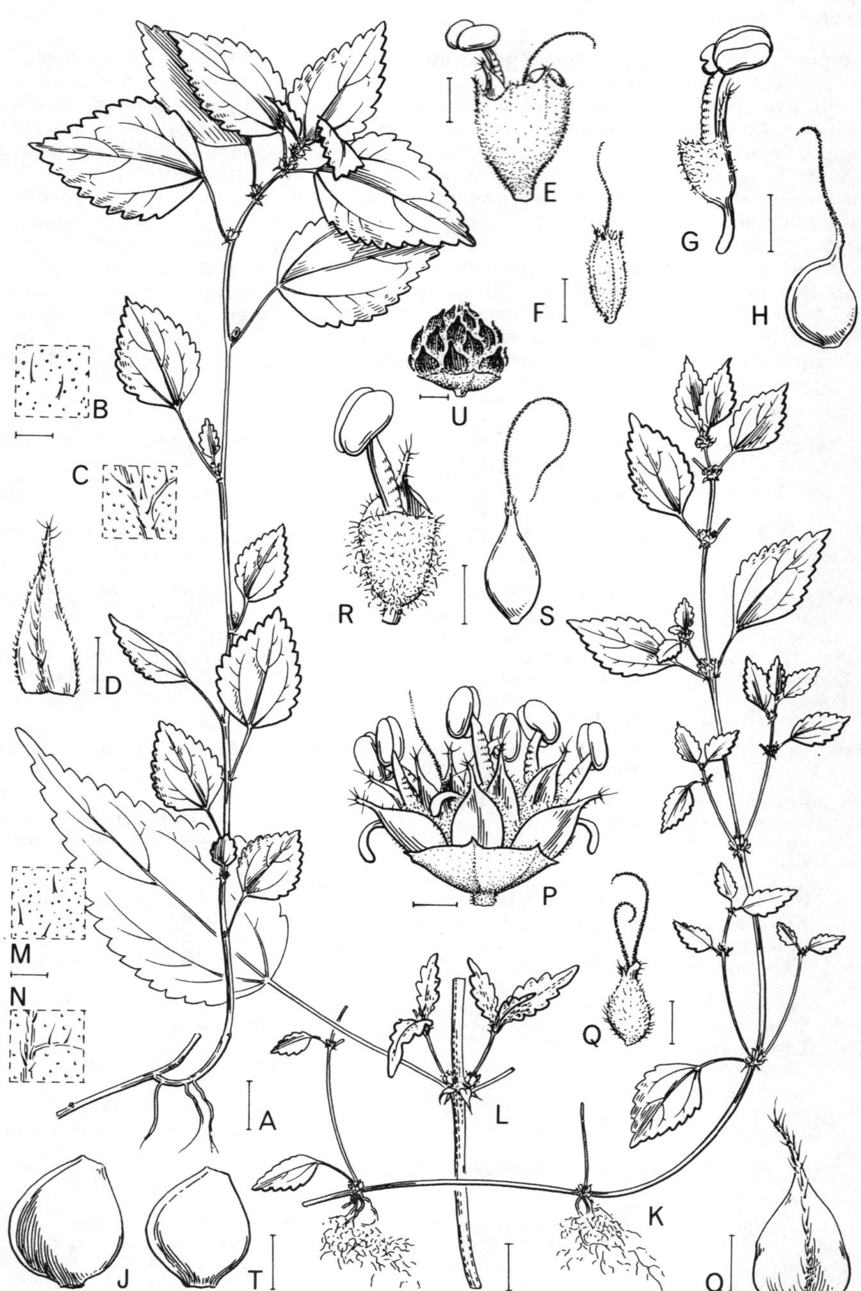

FIG. 16. *DROGUETIA DEBILIS* — **A**, habit; **B**, leaf, upper surface; **C**, lower surface; **D**, stipule; **E**, bisexual involucre, with two male and one female flowers; **F**, female involucre with single flower; **G**, male flower; **H**, female flower; **J**, achene. *D. INERS* subsp. *INERS* — **K**, habit; **L**, node; **M**, leaf, upper surface; **N**, leaf, lower surface; **O**, stipule; **P**, bisexual involucre, some reflexed filaments have lost their anthers, one style visible; **Q**, female involucre with two flowers; **R**, male flower; **S**, female flower; **T**, achene; **U**, young involucre, young male flowers with conspicuous indumentum. Bar lengths: A, K, L = 1 cm.; B–H, M–S = 1 mm.; J, T, U = 0.5 mm. Reproduced from the Nordic Journal of Botany 8: 41 (1988). Drawn by Eleanor Catherine.

Perennial herb with prostrate and ascending stems, up to 25 cm. tall, branching from the base only. Young branches and petioles with sparse, somewhat appressed, coarse stiff hairs. Leaves alternate, except for the lowest 2 pairs of leaves on young plants which are apparently always opposite; stipules linear-lanceolate, up to 2 mm. long and 1.5 mm. wide; petiole up to 2.5 cm. long; lamina ovate, up to 3 cm. long and 2 cm. wide, base cuneate, margin dentate with 7–11 teeth on each side, apex acute, apical tooth longer than wide; lateral nerves 2–3 pairs, basal pair arching towards the midnerve and anastomosing with upper lateral nerves; upper surface with very sparse scattered stiff hairs, lower surface with stiff hairs on the nerves. Inflorescences axillary involucres, 2–5 in the leaf-axils on the upper part of the stem, of two kinds occurring together, the majority bisexual, broadly campanulate up to 3 mm. in diameter with 1 ♀ flower and few–many ♂ flowers, others with only 1–2 ♀ flowers. Fig. 16/A–J.

KENYA. Northern Frontier Province: Mt. Kulal, *Bally* 5597!; Trans-Nzoia District: Cherangani Hills, Kapolet [Kabolet], Aug. 1963, *Tweedie* 2698!; Meru District: Upper Imenti Forest, June 1974, *Faden* 74/905!
TANZANIA. Mbulu District: Mt. Oldeani, Sept. 1961, *Newbould* 5886!; Arusha District: Ngurdoto Crater, Oct. 1965, *Greenway & Kanuri* 11959! & Mt. Meru, Dec. 1966, *Richards* 21828!
DISTR. **K** 1, 3, 4, 6; **T** 2; E. Zaire
HAB. Undergrowth of upland rain-forest and dry montane forest; 1050–2750 m.

NOTE. With two collections from the lake region of NE. Zaire this species must be expected to occur in Uganda.

14. AUSTRALINA

Gaudich. in Freyc., Voy. Monde, Bot.: 505 (1830); G.P. 3: 394 (1880); Engl. in E. & P. Pf. 3(1): 117 (1888); G.F.P. 2: 194 (1967); Friis & Wilmot-Dear in Nordic Journ. Bot. 8: 53 (1988)

Perennial herbs with prostrate and erect or ascending stems; monoecious, but with unisexual inflorescences. Leaves alternate or opposite, petiolate; cystoliths punctiform or linear; stipules lateral, free, or, in case of the leaves being opposite, partly fused to form interpetiolar stipules; lamina serrate or crenate. Bracts absent or minute, at the bases of only the ♂ flowers. Male inflorescences pedunculate in the axils of the upper leaves: ♂ flowers in a whorl at the end of the peduncle, (1–)2–4(–5); perianth conically fused at the base, unequally lobed in the upper part, the tip of the longer lobe inflexed at first, later reflexed, enclosing the single stamen; stamen at first inflexed, later reflexed; rudimentary ovary absent. Female inflorescences usually axillary, usually below the ♂ inflorescences, consisting of a sessile cluster of 1–5 flowers; perianth tubular, with ± 5 obtuse teeth at the apex, surrounding the erect ovary which is terminated by a filiform style. Achene ovate, enclosed in the persistent perianth.

A genus of two species, one in Australia and New Zealand, the other in Ethiopia and Kenya.

A. flaccida (*A. Rich.*) *Wedd.* in DC., Prodr. 16(1): 235/60 (1869); Rendle in F.T.A. 6(2): 305 (1917); U.K.W.F.: 325 (1974); Friis & Wilmot-Dear in Nordic Journ. Bot. 8: 56, fig. 20H–N (1988). Type: Ethiopia, Tigre, between Adua and Memsah, *Quartin Dillon* (P, holo.!)

Presumably perennial herb with erect stems, up to 15 cm. high, with thin, flexible, almost filiform stolons up to 30 cm. long, on which are opposite cataphyls. Younger stems and petioles with stiff erect hairs of two different lengths. Leaves opposite; stipules from a pair of opposite leaves partly fused to form interpetiolar stipules, up to 4 mm. long; petiole very short, up to 1 mm. long at most; lamina narrowly elliptic to lanceolate, 2.5–3.5 cm. long, 1–1.5 cm. wide, base cuneate, margin serrate with (3–)5–9 acute teeth on each side, apex acute, apical tooth up to twice as long as broad; lateral nerves 2–4 pairs, basal pair reaching almost ¾ towards the apex before arching towards the midnerve and anastomosing with upper lateral nerves; both surfaces with sparse to abundant stiff erect hairs. Male inflorescences pedunculate, axillary at the upper leaves; peduncle up to 0.5 cm. long, with indumentum as on the stem, but usually more sparse; ♀ inflorescences sessile in the lower leaf-axils (occasionally 1–a few ♀ flowers found in the axils of cataphyls on the stolon), 1–3 together. Male flowers 2–4(–5), with a minute bract at the base of each; perianth up to 2 mm. long, with some stiff hairs on the outside. Female flowers shortly pedicillate; perianth up to 2.5 mm. long, cylindrical, glabrous on the outside, with several ridges and ± 5 blunt teeth at the apex, between which the filiform style projects. Fig. 17/H–N, p. 61.

KENYA. Trans-Nzoia District: E. Elgon, Elephant's Cave, Aug. 1962, *Tweedie* 2410! & Aug. 1963, *Tweedie* 2677! & Sept. 1966, *Tweedie* 3334!
DISTR. **K** 3; Ethiopia (Tigre, Gondar, Shoa)
HAB. In shade among rocks, or on moist rock faces in upland forest; 2500–2650 m.

SYN. *Pouzolzia flaccida* A. Rich., Tent. Fl. Abyss. 2: 259 (1850)
   Australina schimperiana Wedd. in Ann. Sci. Nat., sér. 4,1: 212 (1854) Type: Ethiopia, Woina, *Schimper* 795 (P, holo.!, G, iso.!)

NOTE. In the Flora area known only from the collections cited.

15. DIDYMODOXA

Wedd. in Archiv. Nat. Hist. Nat. Paris 9 (Monogr. Urtic.): 547 (1856) & in DC., Prodr. 16(1): 235/61 (1869); Friis & Wilmot-Dear in Nordic Journ. Bot. 8: 45 (1988)

Australina sensu auct. mult., *non* Gaudich. (1830)

Annual herbs, monoecious or (by abortion) dioecious. Leaves alternate, petiolate, serrate, crenate or entire; cystoliths punctiform; stipules free, lateral. Inflorescences usually bisexual, axillary, sessile, bracted, bracts free, not prominent, usually shorter than the flowers. Flowers: ♂ with a boat-shaped, sometimes almost bract-like perianth with erect tip and slightly fused cylindrical base, with 1 stamen, without rudimentary ovary; ♀ without perianth, ovary erect, with a subcapitate or shortly linear style. Achene slightly winged along one side, other side rounded, frequently 2 ovaries or achenes join along the rounded side to form a double fruit, which is much larger than two single fruits, both seeds developing in this case. Fig. 17/A–G.

Two species in South Africa and eastern tropical Africa, as far north as N. Ethiopia.
What was formerly known as the genus *Australina*, distributed in Australia and Africa, has recently been divided into two genera by Friis & Wilmot-Dear in Nordic Journ. Bot. 8: 25–59 (1988), who resuscitated Weddell's genus *Didymodoxa*, applying the name to two of the African species, while *Australina* is restricted to two species, one from Australia and New Zealand and the other from Ethiopia and NW. Kenya.

D. caffra (*Thunb.*) *Friis & Wilmot-Dear* in Bol. Soc. Brot., sér. 2, 58: 210 (1986). Lectotype, chosen by N. E. Brown in K.B. 1913: 80 (1913): South Africa, without further locality, *Herb. Thunberg* 22124 (UPS-THUNB, lecto.!)

Slender annual herb. Stems up to 80 cm. high, branched from the base, lower part of branches sometimes rooting at nodes, moderately to sparsely pubescent with patent stiff or slender curly hairs, sometimes also with a few hooked hairs. Lower 1–2 pairs of leaves opposite, upper leaves alternate; stipules lanceolate, long-acuminate, glabrous or ciliate, 2–5 mm. long, 0.3–0.8 mm. wide; petiole 25–47 mm. long, pubescent or with scattered stiff hairs; lamina lanceolate, 2–7.5 cm. long, 1.5–4 cm. wide, base cuneate, margin clearly crenate with 5–12 teeth on each side; lateral nerves 4–5 pairs, basal pair reaching the upper ⅓ of lamina before arching towards the midnerve and uniting with the upper lateral nerves; upper surface with sparse appressed stiff hairs, lower with hairs numerous on veins and a few scattered hairs. Inflorescences up to 5 mm. in diameter, bracts ciliate, linear to narrowly lanceolate. Male and ♀ flowers as described under the genus. Achenes covered with hooked hairs and glands, ± 1.75 mm. long. Fig. 17/A–G.

KENYA. Northern Frontier Province: E. Mt. Kulal, near Gatab, Nov. 1978, *Hepper & Jaeger* 7028!; N. Nyeri District: Zawadi Estate, 7 km. on Nyeri–Kiganjo road, June 1974, *Faden & Evans* 74/682!; Nairobi National Park, May 1961, *Verdcourt & Polhill* 3161!
TANZANIA. Moshi District: Kilimanjaro, Marangu, Jan. 1894, *Volkens* 1700! & Marangu, Dec. 1963, *Archbold* 389! & Mashati, Jan. 1929, *Haarer* 1735!
DISTR. **K** 1, ?3, 4, 5; **T** 2; Ethiopia, ?Zaire/Zambia, Namibia, South Africa (Natal, Orange Free state, E. Cape Province as far south as Port Elizabeth)
HAB. Upland rain-forest and dry montane forest, among rocks or in shaded grassland along streams; 1400–2000 m.

SYN. *Urtica caffra* Thunb., Prodr. Pl. Cap. 1: 31 (1794)
   Australina acuminata Wedd. in Ann. Sci. Nat., sér. 4,1: 212 (1854); Rendle in F.T.A. 6(2): 306 (1917); N. E. Br. in Fl. Cap. 5(2): 555 (1925); Roessler in Prodr. Fl. SW.-Afr., 17, Urticac.: 2 (1967); U.K.W.F.: 325 (1974). Type: South Africa, Natal, Yellowwood R., *Drège*, specimen marked "*Parietaria cuneata* E.M. a" (P, holo.!, G, K, iso.!)

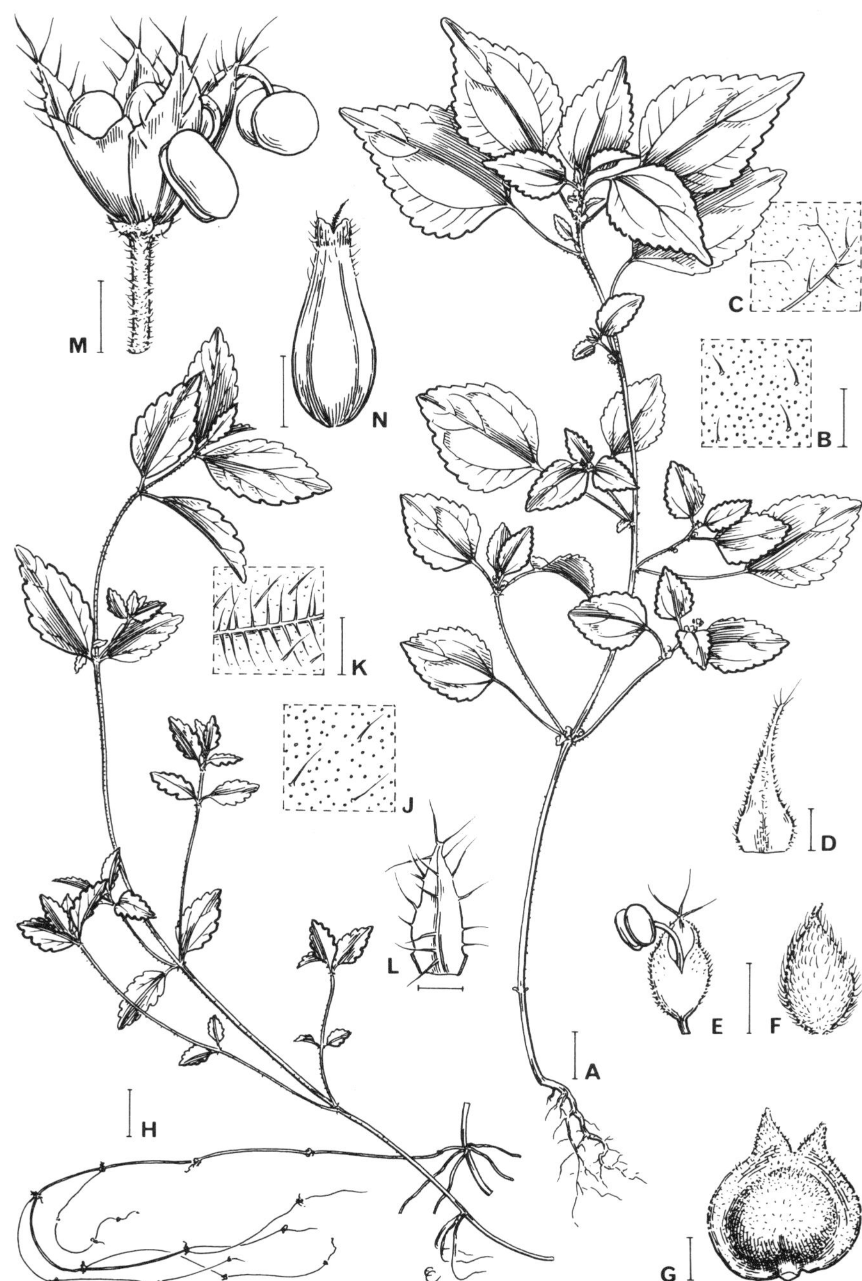

FIG. 17. *DIDYMODOXA CAFFRA* — **A**, habit; **B**, leaf, upper surface of leaf; **C**, leaf, lower surface of leaf; **D**, stipule; **E**, male flower; **F**, female flower, with remains of withered stigma; **G**, "double fruit", with two achenes fused along the rounded, unwinged side. *AUSTRALINA FLACCIDA* — **H**, habit; **J**, leaf, upper surface of leaf; **K**, leaf, lower surface of leaf; **L**, stipule, at base fused with its opposite number; **M**, male inflorescence; **N**, female flower. Bar lengths: A, H = 1 cm.; B–G, J–N = 1 mm. Reproduced from the Nordic Journal of Botany 8: 50 (1988). Drawn by Eleanor Catherine.

Didymodoxa acuminata (Wedd.) Wedd. in Archiv. Mus. Nat. Hist. Nat., Paris 9 (Monogr. Urtic.):
549 (1856)
Didymodoxa cuneata Wedd. in DC., Prodr. 16(1): 235/62 (1869), *nom. superfl.*, based on *A.
acuminata*
Droguetia umbricola Engl., P.O.A. C: 164 (1895). Type: Tanzania, Kilimanjaro, Marangu, *Volkens*
1700 (B, holo.†, BM, iso.!)
Pouzolzia erythraeae Schweinf. in Bull. Herb. Boiss. 4, Appendix 2: 146 (1896); Rendle in F.T.A.
6(2): 294 (1917). Type: Ethiopia, Eritrea, Lava valley, *Schweinfurth* 1658 (B, holo.†)
Australina caffra (Thunb.) Prain in Ann. Bot. 27: 388 (1913)
Pouzolzia piscicelliana Buscalioni & Muschler in E.J. 49: 465 (1913); Piscicelli, Nella reghione dei
Laghi Equatoriali: 106, t. (1913); Rendle in F.T.A. 6(2): 294 (1917). Type: Zambia/Zaire,
Bwana Mukuba–Sekontui, 1200 m., *Aosta* 512 (B, holo.†)

INDEX TO URTICACEAE

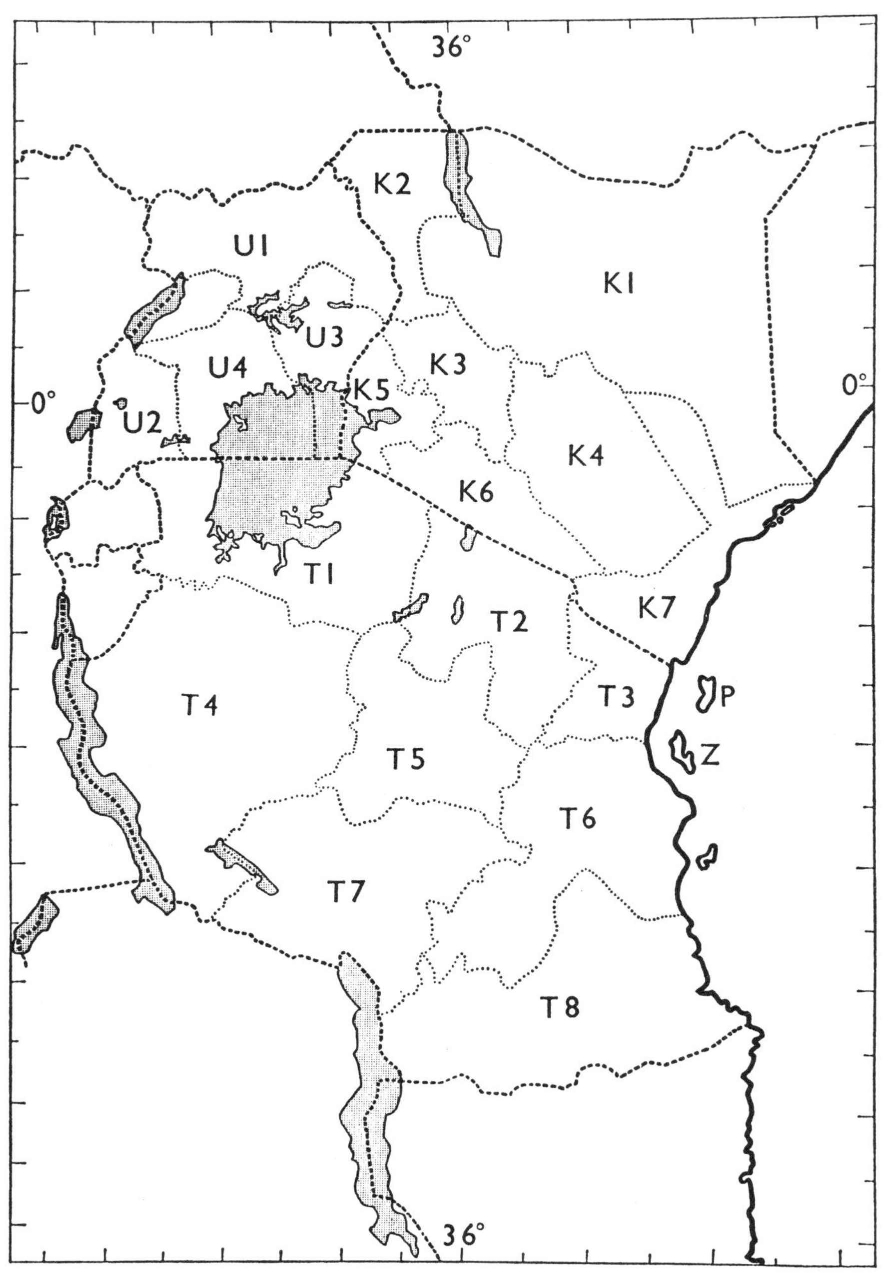
36°
K2
U1
K1
U3
U4
K3
K5
0°
U2
K4
0°
K6
T1
K7
T2
T3
P
T4
Z
T5
T6
T7
T8
36°